EcoProduction

Environmental Issues in Logistics and Manufacturing

Series Editor

Paulina Golinska-Dawson, Poznań, Poland

The EcoProduction Series is a forum for presenting emerging environmental issues in Logistics and Manufacturing. Its main objective is a multidisciplinary approach to link the scientific activities in various manufacturing and logistics fields with the sustainability research. It encompasses topical monographs and selected conference proceedings, authored or edited by leading experts as well as by promising young scientists. The Series aims to provide the impulse for new ideas by reporting on the state-of-the-art and motivating for the future development of sustainable manufacturing systems, environmentally conscious operations management and reverse or closed loop logistics.

It aims to bring together academic, industry and government personnel from various countries to present and discuss the challenges for implementation of sustainable policy in the field of production and logistics.

More information about this series at http://www.springer.com/series/10152

Arpita Saxena
Editor

Biotechnology Business - Concept to Delivery

Springer

Editor
Arpita Saxena
SPAK megAcorp
Aurangabad, India

ISSN 2193-4614 ISSN 2193-4622 (electronic)
EcoProduction
ISBN 978-3-030-36129-7 ISBN 978-3-030-36130-3 (eBook)
https://doi.org/10.1007/978-3-030-36130-3

This Springer imprint is published by the registered company Springer Nature Switzerland AG
The registered company address is: Gewerbestrasse 11, 6330 Cham, Switzerland

This work is dedicated to ***selfless efforts*** *of* ***Mentors****, who, with their expertise and experience nurture young start-ups and entrepreneurs throughout the globe.*

Foreword

It gives me pleasure to write a foreword for the book compiled with the basic rationale of culturing the next generation with an enterprising attitude. As a leader of synthetic biology in one of the leading business houses, I have addressed various challenges to sustainable business solutions. In all this, I see the drive to bring about the mental accommodation of an individual to channel towards a new idea. To come out of one's comfort zone and strive for excellence is desired by every fresh mind but when it comes to practically doing it, things are pushed towards doubt and drift. It needs a good mentor and a strong sense of belonging to the idea to actually reach the zenith.

This book provides clear landmarks which any start-up would come across. Right from the beginning of the book till the end, the process of conceiving a business idea to delivering it in the form of a product is discussed. To give it more appealing, the parts of an entrepreneur's journey are projected in the form of musical scales. To begin with, the first chapter details from the beginning of biotechnology up to its contemporary extent. Research students' favourite topic is covered in the next part, where the molecular, genetic or nano targets, the ones used to manipulate the working pathways at various levels, are discussed. The book in the following chapters covers nutraceuticals and their importance in health care and biofuels and their contribution to the environment.

The chapters in the following parts cover other pertinent aspects like regulations, IP, commerce and management of a biotech start-up. These are the important aspects which are to be precisely known to individuals when they start on their own.

The knowledge and contribution of the authors of the book are commendable as they have played the role of mentors for the readers. I would also congratulate the editor of the book Dr. Arpita Saxena for the idea and effort put in the book. I hope this book is used well by the young students to earn experience and know-how to start-up and proceed with interesting and useful ideas. Wishing the best to the entire team related to this book.

July 2019

Dr. Santanu Dasgupta
Senior Vice President
Synthetic Biology
Reliance Industries Limited
Mumbai, India

Preface

The first thing that comes to our minds when we read the title of this book is—what is the business of biotechnology, or business of anything which is a discipline itself? How can one approach towards a subject as a profit or loss statement? And many more thoughts that cross our brains, which are by the way a symbol of our affection, respect and sense of gratitude towards the subject…

So, how did I think about it as a title for my next book? This opens up the Pandora's box of my past experiences and story of how this idea looks genuinely graceful and useful to me. Few days back, a self-proclaimed enterprising professor in one of his addresses to the candidates who were attending an entrepreneurial skill development program in one of the university incubation centres said, 'if I give you a thousand bucks to sell vegetables for one day, who amongst you would like to do it?' and before anybody could raise their hands, he himself answered his question as 'obviously none of you… since you are all Ph.D.s, and the task will not give you job satisfaction'.

This was the time when I had left my job from a start-up company to have my own venture, and I was struggling to pay my bills. The first thought that crossed my mind was—what actually gives job satisfaction? The ability to be on your own or the perfection at whatever you do? Or maybe growth which gives you a boost at work or an environment that keeps you charged and motivated? If all this counts, what is wrong with selling vegetables without any monetary investment and making thousand bucks a day? The answer may be different for different people and that is what defines our approach to our subjects. To me, being able to tap the smallest of opportunity and trying to solve the most complex problems in simplistic ways are the attitude of an entrepreneur with fair chances of success.

The day left me reflecting on one more point and that was, if somehow the next generation is told that there are no high and low profiles in work when you own your business, and that challenges have to be met with wit and zeal, they might be better prepared for things to come. Even better is to discuss the subject w.r.t. application, discuss real challenges that confront them to realize their goals and tell them real stories of success as well as failures of entrepreneurs. This book is an attempt in the same direction although this also might have some imperfections, but the sincerity in efforts is promised.

Biotechnology, as the name has it, is a beautiful orchestration of multiple technologies working with living cells. These natural and human intervened combinations of A-T, G-C bring about lot of products and alternatives to combat the challenges in health care, agriculture, nutraceuticals as well as environmental protection. This book is a next step towards strengthening the entrepreneurial environment in the field of biotechnology specifically in India. Rhythm is the basis of life. So whether it is the lub-dub rhythm of a living heart or the rhythm of the life of an entrepreneur, efforts have to be continuous, dedicated and enriched with knowledge and experience. Here, we bring every important aspect of a biotech enterprise in the form of **Musical scales**. Step by step as the young biotechnologists walk this road, they may be able to compose their own songs…

The first part of the book **Do**—'Why biotechnology' draws attention of people about what to expect from biotechnology as a discipline. It flows from past of human curiosities and discoveries to present scope of their applications. Here, an insight into the scope and depth of the biotechnology as a subject is provided to the readers.

The next is the most interesting part **Re**—'Tiny targets big impact'. This part will include the innovations through targeting at nano/genetic/molecular levels of cells, to better the existing systems of health care, crop production/protection, etc. The revolutionary CRISPR-Cas9 genome editing tool will be discussed along with its applications. Another chapter in the same category discusses the role of nanotechnology in increasing the efficiency of delivering new forms of therapeutics to the target tissue. This field is the newest and most exciting, hence an important include in the book.

Mi—Food for all; Nutraceuticals 'The intellectual quests only start when the tummy is full'. Ensuring nutrition for all, this segment describes various approaches in nutrition management and nutraceuticals production, scope which play a huge role in disease prevention and general health promotion. As we talk about entrepreneurship, it is important to understand the demand and supply chain. The food and nutrition market especially those targeting specific groups like infants, children, pregnant women and patients is immense, and a newcomer needs to realize the need and choice of his/her product.

Fa—Biofuels—Recently, biofuels have surfaced as alternative source of energy which are sources from plants or specifically algae. Considering the limitations of first and second generation of biofuels, the third generation of biofuels has attracted the research communities. Microalgae is striving to establish its place globally as an alternative to conventional fuels. Necessity of fuels is obvious and with the contemporary fuels nearing extinction, the world looks forward to alternatives. Biofuels, being one of those alternatives, makes an important placeholder in research as well as this book.

So—Regulations—Every production is monitored with certain set of rules and regulatory bodies to ensure safety and quality. This monitoring becomes a prerequisite when the modifications at DNA or molecular level are concerned and that why regulations part will be discussed in this segment of book. Once an entrepreneur is ready with an idea of product, it is indispensable to know the rules and restrictions for its production.

La—Intellectual Rights. The production is not always in terms of materials but also in terms of intellectual assets. Here, the author throws light on the intellectual property rights as only material wealth is no more all that a person has, and it is important for a start-up to understand the legalities of registration of assets as well as ways to protect them.

Ti—Commerce and Management. Latest trends in biotechnology market and various approaches to enter into it will be the focus of this segment. Once the budding researcher turns a budding entrepreneur, he/she needs management skills specific to a biotech industry. This segment gives insights into this aspect.

Products of modern biotechnology are discussed in the end to motivate the readers. It also gives them an exposure about the line of products available already, in the process of production or the ones which are not thought of still.

This book will serve the undergraduates and graduates with the guidance to prepare themselves with entrepreneurial skills and mindsets, so that when they are at the verge of completion of their courses, they are capable of joining big industries and/or start their own start-ups.

This book is a guide through the various aspects of biotechnology as a subject and the opportunities it offers to the coming scientific community. The authors of the book belong to various specialized sectors from all around the globe. They have a technique of changing their most significant research into simple, understandable and lucid English. Their contributions bring on the table keen research ideas demanding a common man's attention.

The impact of this book would be long imprinted in the minds of the readers. Wishing you a happy read.

Aurangabad, India — Arpita Saxena

Contents

Regulations

Intellectual Rights

Commerce and Management

Some Success Stories and Products

Editor and Contributors

About the Editor

Arpita Saxena is an entrepreneur, heading a start-up 'SPAK megAcorp' an analytical services and training provider based in Aurangabad, Maharashtra, India. She is striving to better the education system in India with her concept of 'AAPT', which is a social venture of 'SPAK megAcorp'. Arpita received her Ph.D. in biotechnology in 2012 off campus from Guru Nanak Dev University, Amritsar, India, while working at the department of Cancer Pharmacology, Indian Institute of Integrative Medicine (CSIR), Jammu, India. During Ph.D., her work was focused on understanding the methods to evaluate cell death, apoptosis and the mechanisms underlying apoptosis. While during screening a new synthetic, semisynthetic or natural entity for its potential as an anticancer agent on in vitro, in vivo, mechanistic, acute toxicity models, her specific topic was 'Anticancer potential of Spiro derivatives of Parthenin (a weed in Indian Subcontinent) against leukaemia cell lines'.

Later, she was working with a start-up at Aurangabad, funded by BIRAC, India, for 2 years where she was making targeted therapies for blood cancer. She also got an exposure of material science practices during this therapeutic development process. Arpita is associated with BYST (a not-for-profit organization to encourage entrepreneurship amongst Indian youth). Teaching and writing continue to be her motivation.

Contributors

Yashbir S. Bedi Indian Institute of Integrative Medicine, Jammu, India

Ranjana Bhati Department of Microbiology, Bundelkhand University, Jhansi, Uttar Pradesh, India

Sohinee Bhattacharyya Oregon Health and Science University, Portland, USA

Sumit G. Gandhi Plant Biotechnology Division, CSIR-Indian Institute of Integrative Medicine, Jammu, India;
Indian Institute of Integrative Medicine, Canal Road, Jammu, India

Satish Kumar Gupta Indian Institute of Integrative Medicine, Jammu, India

Vijay Lakshmi Jamwal Plant Biotechnology Division, CSIR-Indian Institute of Integrative Medicine, Jammu, India

Nitika Kapoor Plant Biotechnology Division, CSIR-Indian Institute of Integrative Medicine, Jammu, India

Inshad Ali Khan Indian Institute of Integrative Medicine, Jammu, India

Harsh Kumar Regional Centre for Biotechnology, NCR Biotech Science Cluster, Faridabad, Haryana, India;
Manipal Academy of Higher Education, Manipal, Karnataka, India

Durga Prasad Mindala Indian Institute of Integrative Medicine, Jammu, India

Anindit Mukherjee Oregon State University, Corvallis, USA

Srividhya Ravichandran Department of Biotechnology, Indian Institute of Technology, Chennai, India

Arpita Saxena AAPT Foundation, SPAK megAcorp, Aurangabad, India

Manish R. Shukla Research and Development Center, Reliance Industries Limited, Navi Mumbai, India

Gaurav Verma Clinical Research Center, Lunds University, Malmo, Sweden

Tabassum Zafar Department of Bioscience, Barkatullah University, Bhopal, India

Why Biotechnology?

Biotechnology: Discoveries and Their Applications in Societal Welfare

Harsh Kumar

Abstract The present chapter describes the field of biotechnology from earliest known mention till modern-day developments. The introduction throws light on understanding the practice of using microorganisms for making food since prehistoric times and covers aspects on key discoveries in recent past like vaccines and drugs. The section journeys through important seminal contributions that laid the foundation of many branches of biotechnology which would eventually be applied for the benefit of the population. The gradual advancement in the field of knowledge about life, its complexities and processes governing these led to the generation of medicines, diagnostic tests, industrially important materials and environmentally sustainable products in future years. Biotechnology can be classified based on the broad spectrum of deliverables it caters to the society. Starting with the healthcare, which constitutes red biotechnology, the text details significant discoveries and inventions that have greatly enhanced biomedical research. Green biotechnology, mentions important methods which have been applied in the field of crop and livestock improvement to ensure food security. Blue biotechnology, an emerging area enlists and highlights important marine genetic resources which are being adapted for various demands of global economy. Finally, the last section details the application of biotechnology in the field of industry and environment for the generation of better raw materials and its clean-up, respectively.

Keywords Biotechnology · Health care · Gene · Agriculture · GM · Marine · Drugs · Environment · Industry

H. Kumar (✉)
Regional Centre for Biotechnology, NCR Biotech Science Cluster, Faridabad, Haryana 121001, India
e-mail: visitharsh@gmail.com

Manipal Academy of Higher Education, Manipal, Karnataka 576104, India

A. Saxena (ed.), *Biotechnology Business - Concept to Delivery*, EcoProduction,
https://doi.org/10.1007/978-3-030-36130-3_1

1 Introduction

> Biotechnology is the integrated use of biochemistry, microbiology, and engineering sciences in order to achieve technological (industrial) application of the capabilities of microorganisms, cultured tissue cells and parts thereof. (European Federation of Biotechnology 1981)
>
> Biotechnology means any technological application that uses biological systems, living organisms, or derivatives thereof, to make or modify products or processes for specific use. (Convention on Biological Diversity, United Nations 1992)

The term biotechnology has been defined in different ways by people and organizations. Scientific community by and large agrees that the science at its core involves the usage of other life forms for human betterment. Since time immemorial, people across various cultures have been using living organisms as food or medicine either in their natural or processed forms. These included whole microbes or parts of plants, animals or their derived products. Processing of raw food items required more elaborate methods which employed whole organisms, e.g. yeast. The technique of fermentation in preparation of edible items (e.g. bread, cheese, curd, wine, etc.) was a traditional knowledge across various civilizations. Since 6000 B.C., herdsmen of Central Asia knew the art of curd/yogurt making by carrying the milk in animal stomachs. Around the same time, the people of Babylon and Sumer (present-day Iraq) had the know-how of wine making by utilizing yeast. During 4000 B.C., people of ancient Egypt employed yeast in bread making (Bud 1993). Chinese in 500 B.C. used poultices made up of curds having growth of moulds in it as a source of antibiotics for curing boils, and farmers in Ukraine applied cheese moulds on wounds to cure infection. At around 470 B.C., Greek philosopher Socrates expounded the observation of shared features between the parents and their offspring. Another Greek philosopher Aristotle in around 300 B.C. stated that children inherit traits from their father (Daston and Lunbeck 2011).

Life in primitive times was principally dependent on agriculture and livestock. Man began collecting wild plants for consumption since 7000 B.C. First organized method of agriculture dates back to 3000 B.C. in China where crop rotation was practiced. Since then, man started observing plants closely. Theophrastus (322 B.C.) mentioned about the diseases affecting crop plants and pointed out bad air and nutrition as their cause (Singh 2018). Domestication of crop plants was a significant achievement as it ensured food security by preserving the seeds for next seasonal round of sowing. Animals that were domesticated for the first time were mainly cattle, sheep and goats as they helped in the cultivation of crops and also provided milk and meat. To carry forward the generation of animal, people began paying attention to animal breeding methods whereby they learned simple mating process that helped in propagation of livestock. Although, people at that time were unaware of the causative agent(s) behind these processes, yet they possessed experiential knowledge. As the thought process evolved, people began to provide a more rational explanation for biological phenomena they observed. Centuries later, from 1700 A.D. onwards numerous milestone discoveries were made that led to the improvement in our understanding of heredity, causes and cures of diseases, better agricultural practices, etc. Important contributions in the field of basic biology expanded the horizon of knowledge leading to identification of their translational potential (Fig. 1). Many of the older practiced methods were better explained later when people began scientific quest to Understand natural processes. For example, fermentation was explained by Louis Pasteur who in multiple experiments demonstrated bacteria caused souring of milk due to lactic acid formation (Pasteur 1879). Around the same time, Edward Jenner's discovery of smallpox vaccine was based on the observation of acquired immunity and heralded a new era of immunization-based preventive medicine (Riedel 2005). The curing of the wound infection was achieved through curd or cheese moulds in prehistoric times without knowing that poultices made up of them contained antibiotics. Alexander Fleming (1928) purified penicillin antibiotic from the moulds and demonstrated its curing abilities of bacterial infections. The discovery is considered to be one of the greatest scientific achievements in post World War II era as it was able to save countless lives (Aldridge 1999).

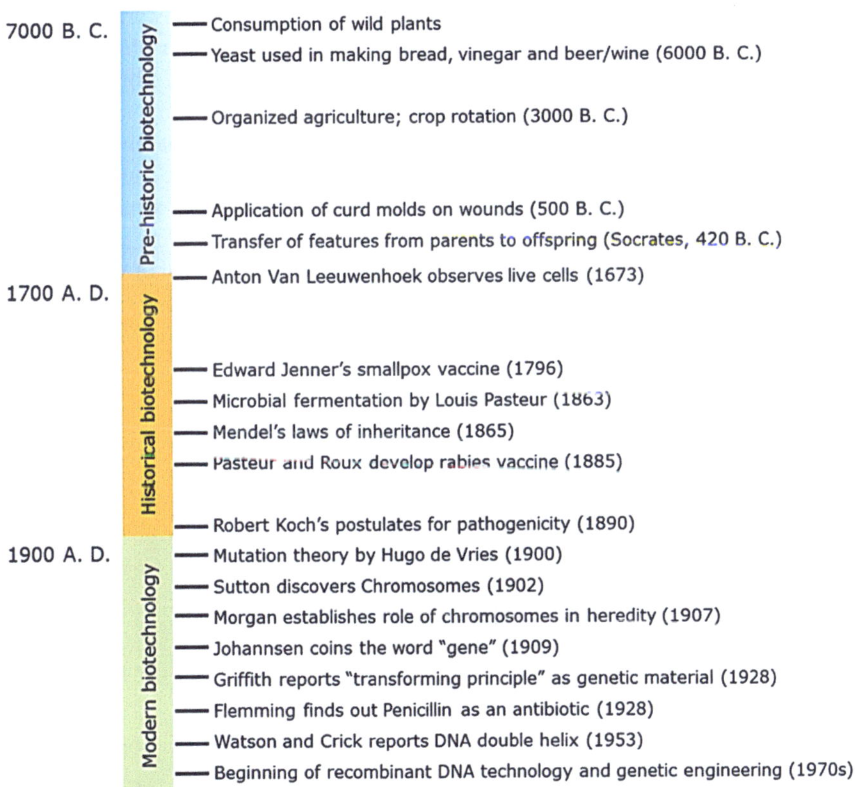

Fig. 1 Brief timeline of events in the development of biotechnology

2 Historical Biotechnology—1700s Onwards

This section describes selected milestone inventions and discoveries that influenced the society in the much needed way in those times. Hunger and disease have been the major challenges before the society since ever. The population was experiencing fatalities caused by several infectious diseases during the eighteenth and nineteenth centuries, e.g. smallpox, tuberculosis, rabies, etc. For most of the diseases, their causative agents were unknown at that time. Identification of disease pathogens and preventive immunization were major scientific breakthroughs of those times which greatly improved public health and increased life expectancy.

2.1 *Smallpox Vaccine*

Smallpox was the deadliest of all diseases of classical times that had the potential to decimate vast majority of population as can be read in history. Edward Jenner, an English physician, got the idea of developing a crude vaccine after observing that milkmaids working on a cow farm were immune to smallpox virus. This made him reason that may be the exposure of milkmaids to cowpox made them resistant to the smallpox virus and therefore such a challenge to a healthy individual may lead to a similar outcome. Jenner liked to experiment this idea for which he inoculated pus taken from a milkmaid suffering from cowpox into the arm of a boy named James Phipps who was found immune many days later. The observations were submitted to the Royal Society in 1797 (Plotkin 2014). Jenner established the field of vaccinology which has now become one of the exciting areas of biotechnology research with great commercial potential.

Edward Jenner. Photo Courtesy of the History of Medicine Division at the U.S. National Library of Medicine

2.2 *Rabies Vaccine and Germ Theory of Disease*

In 1885, another vaccine, this time against rabies was developed by Louis Pasteur and Emile Roux. Rabies developed as a result of bites from the dogs that were carrying rabies virus in their saliva. The virus is known to attack central nervous system that triggers encephalitis and kills the infected person. The vaccine was a great boon for the society because until then all rabies infections led to death (Hicks et al. 2012). The modern understanding of diseases came in the late nineteenth century from the studies done by Louis Pasteur and Robert Koch. Pasteur contributed to the development of "germ theory of diseases" which Koch later extended through his findings. Germ includes any microscopic pathogen like bacteria, virus and protists. The theory states that infectious diseases are caused by the growth and multiplication of germs after they invade human or animal body. Pasteur discovered the theory while examining a patient with puerperal fever whose blood was infected with pyogenic vibrio and suggested the use of boric acid as a potent and safe antiseptic to prevent the growth of germs (Ernst 2014). Germ theory of disease is still acceptable in medical sciences and has greatly improved the understanding of disease aetiology which has led to many preventive strategies and therapeutic interventions.

2.3 *Koch's Postulates*

Dr. Robert Koch, a German physician, in 1882, put forth his famous theory of establishing a link between the disease and the pathogen. He synthesized four important parameters from his studies on tuberculosis (TB) that became gold standard for judging a pathogen for its role in any disease. It must be noted that contemporary research on tuberculosis was unable to identify pathogen causing the disease, and Koch's postulates came out from the studies in which Robert Koch systematically showed the presence of the stained bacillus in the tuberculous tissue. Following are the Koch's postulates:

(1) The microorganism must be found in abundance in all organisms suffering from the disease, but should not be found in healthy organisms.
(2) The microorganism must be isolated from a diseased organism and grown in pure culture.
(3) The cultured microorganism should cause disease when introduced into a healthy organism.
(4) The microorganism must be reisolated from the inoculated, diseased experimental host and identified as being identical to the original specific causative agent.

Until now, TB was wreaking havoc leading to mass deaths due to its spread; therefore, in the light of Koch's findings, public was provided better treatment and care. TB sanatoria were built where the patients were kept in isolation to prevent further transmission (Walker et al. 2006). Koch was awarded Nobel Prize for Medicine and Physiology in 1905 in recognition of his contribution.

2.4 Fermentation Theory

Besides working in the disease biology, Pasteur also demonstrated for the first time the scientific basis of microbial fermentation process by propounding "fermentation theory" (1857). The theory rejected the idea of *spontaneous creation* and was later itself replaced by the germ theory of disease. It was perhaps the only scientific discoveries of its times that had the potential of industrial scalability. Through systematic experiments, Pasteur showed the role of yeast in fermentation process of ethanol production while also giving terms like "aerobic and anaerobic" (Barnett 2003). Fermentation till today continues to be one of the major areas of application of biotechnology on which many food, chemical and drug industries thrive. There are numerous health benefits of consuming fermented food products. Besides improving digestion, fermented edible items have also been shown to boost immunity and relieve from stress. Fermentation is also used to manufacture important industrial chemicals like ethanol, acetic acid; enzymes like cellulase for use in paper and pulp industry; antibiotic like penicillin by culturing fungus *Penicillium notatum* on a large scale (Parvez et al. 2006).

2.5 Penicillin Discovery

Professor Alexander Flemming (1928), a Scottish physician is credited with the discovery of first true antibiotic penicillin having a broad spectrum bactericidal activity. The antibiotic was discovered as an observation of inhibited bacterial growth on a plate having mould (*P. notatum*) growing. The "mould juice" was the antibiotic secreted out by the fungus which was effective in killing wide range of bacteria like *Staphylococcus* (causes abscess), *Meningococcus* (causes meningitis and sepsis), *Streptococcus* (causes pneumonia) and *diphtheria bacilli* (causes diphtheria) (Fig. 2). The antibiotic proved to be a massive life saver in post World War II era to treat wounded soldiers and general public. An otherwise wound infection that led to the development of abscess and subsequent gangrene could now be effectively treated and cured (Gaynes 2017). Fleming eventually received Nobel Prize for Physiology or Medicine in 1945.

Fig. 2 Schematic showing inhibitory effect of *penicillium* mould on growing bacteria

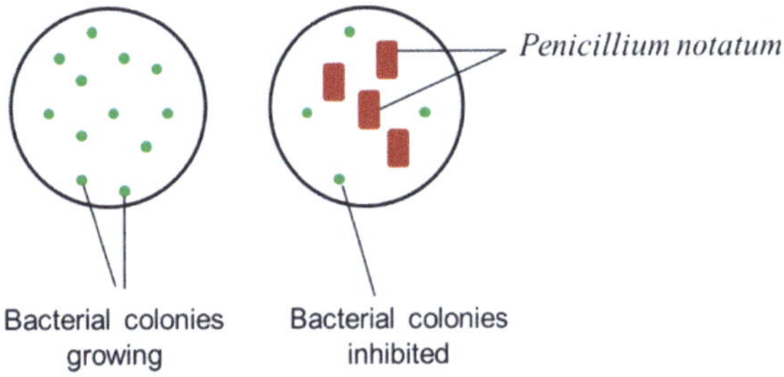

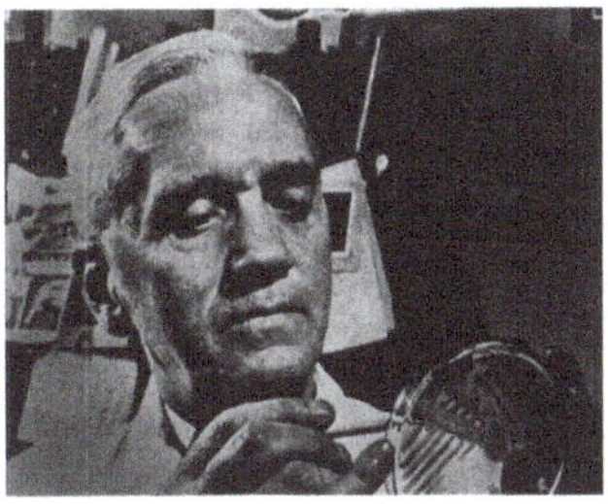

Prof. Alexander Fleming. Photo Courtesy of the History of Medicine Division at the U.S. National Library of Medicine

2.6 Cell Theory and Laws of Heredity

Observing cells for the first time under a microscope was a major breakthrough that totally revolutionized biological research in a way that was going to impact human life in years to come. Zacharias Jansen and his father Hans are credited with inventing the first compound microscope in 1590 (Helden 2010). It was Antonie Philips van Leeuwenhoek who for the first time saw living cells: bacteria, protists, blood cells rotifers, etc. (Lane 2015). Thereafter, some of the most path-breaking reports were published at successive intervals that included the discovery of nucleus by Robert Brown (1831), formulation of "cell theory" by Matthias Schleiden and Theodore Schwann (1839) and addition of new dimension to cell theory by stating "*Omnis cellula e cellula*: all cells arise from pre-existing cells"; by Rudolf Ludwig Carl Virchow (published in *Cellular Pathology*, 1858) that greatly enhanced our understanding of basic functioning of cellular physiology (Kuiper 2010).

Contemporary discoveries of laws of heredity/inheritance by Gregor Johann Mendel (1866; considered to be father of modern genetics) helped in understanding the transmission of traits from one generation to next. The work of Mendel was largely rediscovered by Hugo de Vries, Erich von Tschermak, Carl Correns and William Jasper Spillman. Through simple crossing experiments performed in garden pea (*Pisum sativum*) plant, Mendel was able to conclude the presence of "dominant" and "recessive" traits. What he observed was, when two pure-bred varieties of pea plant (having tall and short traits) were crossed, then the second generation progeny had a mix of population: two were tall and short and remaining two were hybrids (Gros 1992). Initially, Mendel designated the traits as "factors"; however, it was Wilhelm Johannsen (1909) who introduced the term "gene" and William Bateson, the word "genetics" (1905) (Gerstein et al. 2007).

2.7 The Term "Biotechnology"

Karl Ereky, a Hungarian agriculture engineer coined the term *biotechnologie* in 1919 to explain the biological method of converting raw materials into useful products. He wrote a book titled "*Biotechnology of Meat, Fat and Milk Production in an Agricultural Large-Scale Farm*" in which he expressed his views on solving the problem of food crisis (Fiechter 2000).

2.8 Identification of DNA as Genetic Material

Present-day biotechnology revolves around DNA/RNA or protein-based science and its uses. However, to even make use of these molecules for societal application, a detailed and thorough knowledge about them was lacking in the first half of the twentieth century. Frederick Griffith, a British bacteriologist, in 1928 demonstrated the importance of *transforming principle* or the genetic material for the process of bacterial transformation (Lorenz and Wackernagel 1994). Later, in 1944, Avery, McLeod and McCarty showed DNA to be the real transforming principle. In 1953, James D. Watson and Francis H. C. Crick showed the double helix structure of deoxyribonucleic acid (DNA) for which they received the Nobel Prize in Medicine or Physiology (1962) (Crick and Watson 1954; Daston and Lunbeck 2011). With the basic knowledge of DNA, further deep studies in molecular biology became possible which led to the foundation of recombinant DNA technology and genetic engineering.

3 Modern Biotechnology—1970s Onwards

Nirenberg and Matthaei (1961) are credited with the discovery of the genetic code which helped in revealing the information contained in the genes. The triplet arrangement of bases in the mRNA (codons) is matched with the respective anticodons present on the tRNA which carries particular amino acid. During the process of mRNA translation, the codons direct the amino acids they encode to be incorporated in the growing nascent polypeptide chain. The genetic code was further completed and presented as a table by the efforts of Har Gobind Khorana and Robert Holley. The code was found to be degenerated (64 codons encoding 20 amino acids) which meant simply that many codons encode the same amino acid. Nirenberg, Khorana and Holley shared the Nobel Prize in Physiology or Medicine in 1968 (Landmarks 2009).

Werner Arber and Matthew Messelson (1962) discovered "restriction endonucleases" or restriction enzymes that possessed the property of recognizing certain inverted repeat sequences (palindromic sequences) in a plasmid and digest it specifically. The plasmid was a covalently closed circular form of DNA that contained sites for many restriction enzymes called multiple cloning sites (MCS) which allowed incorporation of any gene. Nathans (1971) demonstrated the digestion of the phage DNA of simian virus 40 and resolving of the digested fragments by gel electrophoresis. These findings paved the new exciting area to work with DNA molecule and tune it to do molecular cloning that involved creating an exact replica of a gene (gene cloning). By this time, the central dogma of molecular biology was already in place that dictated the synthesis of protein to be guided by the sequence contained in the gene. People then began exploring the feasibility of cloning the genes that encoded for necessary proteins and began expressing them in suitable host systems (Arber and Linn 1969; Danna and Nathans 1971). The clone (plasmid carrying the gene of interest) created in this manner was called a *recombinant*. The work extended towards the production of recombinantly expressed proteins on an industrial scale, and thus several agricultural, pharmaceutical and other companies jumped into the tray of commercialization, patenting, licensing and mass production of them.

3.1 Hybridoma Technology

Cesar Milstein and Georges J. F. Kohler (1975) invented the hybridoma cells that were capable of synthesizing monoclonal antibodies (mAbs). The purification of single epitope recognizing antibody from the vast repertoire produced by the immune cells had been a challenge before the investigators. Milstein and Kohler fused a plasma B cell (secretes antibodies) with the myeloma cell (cancerous B cells) with the help of sendai virus and created what is called a hybridoma that secreted antibody recognizing single epitope. The antibody thus produced was monoclonal unlike the conventional polyclonal ones that recognized several epitopes (Fig. 3). The hybridoma was transplanted in mice peritoneum where it led to tumour growth and secretion of vast amount of mAb in ascites fluid (Milstein 1999). Milstein and Kohler got Nobel Prize for Medicine and Physiology in the year 1984 for the revolutionizing impact their study had on immunology. With the availability of mAbs, it became possible to selectively identify many tumour cells possessing unique antigens, their separation and subsequent purification. Today, mAbs are used extensively in biomedical and biotechnology research for immunodiagnostics of cancer and its immunotherapy;

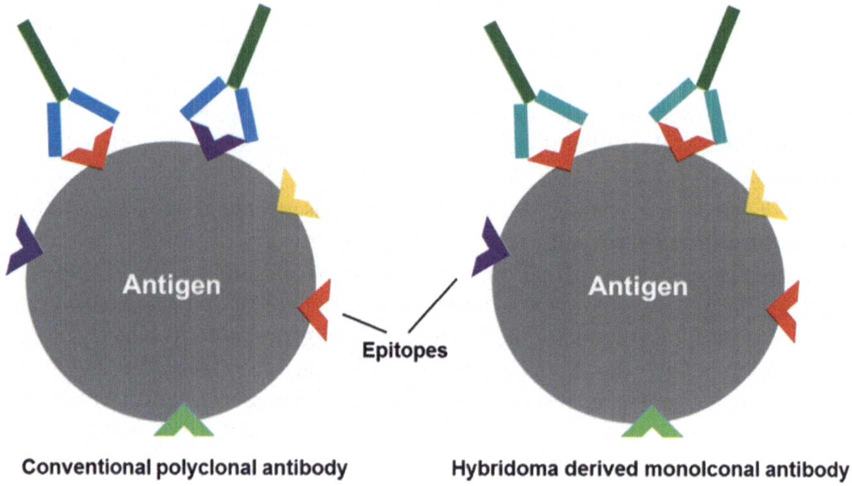

Fig. 3 Schematic illustrating the multiple versus single epitope recognition by a polyclonal antibody and hybridoma-derived monoclonal antibody, respectively

identification of infectious pathogens; affinity purification of single protein from a complex mixture, etc.

3.2 *Insulin: The First Commercial Recombinant Product*

Insulin is a hormone which is responsible for glucose metabolism in the body. Impaired levels of insulin are a hallmark of type 1 diabetes (T1D). The beta cells of islet of langerhans of pancreas produce insufficient or no insulin due to the autoimmune triggered death. Consequential to which the person affected manifests hyperglycaemia (increased blood sugar levels) symptoms. The patients of T1D are administered insulin by subcutaneous injection (Chiang et al. 2014; CDC 2017). The number of diabetics has increased from 108 million (in 1980) to 422 million (in 2014). It is one of the major causes of renal failure, blindness, cardiac attacks, strokes, etc. (WHO 2018). Herbert Boyer and Robert A. Swanson together founded Genentech, Inc. (headquartered in California, USA; now a subsidiary of Roche) in 1976 which became the first biotechnology-based company. Boyer and Swanson entered into the contract agreement to manufacture recombinant human insulin that was until now obtained from pigs. Boyer took the responsibility of scientific aspects of design, execution, purification, etc., while Swanson took care of the finances and legal aspects. The company came out with recombinant human insulin (expressed in and purified from *Escherichia coli*; Fig. 4) in 1978. It was licensed to Eli Lilly and Co. (to be sold as *Humulin*) and was approved by Food and Drug Administration, USA, in 1982 (US F&DA 1982; Hughes 2011).

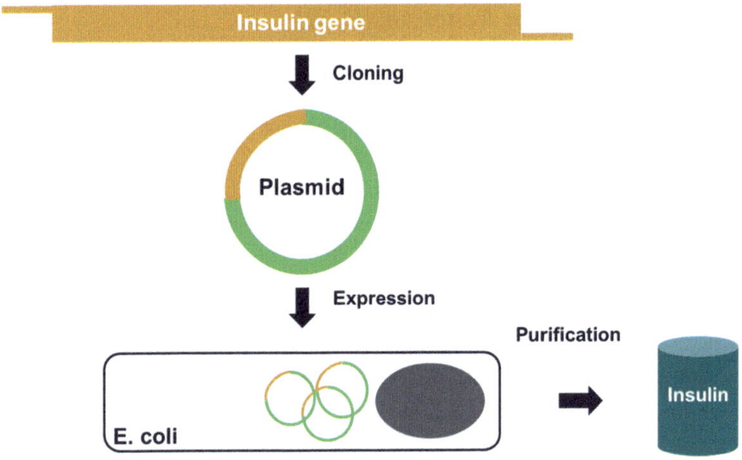

Fig. 4 Schematic depicting the workflow of recombinant human insulin production

Since the original production by Eli Lilly and Co. (Indiana, US); Sanofi (Paris, France) and NovoNordisk (Copenhagen, Denmark) have also come up with their recombinant insulin products (named Insuman and Novolin, respectively) (Landgraf and Sandow 2016). Today, insulin and its derivatives enjoy global sales of over $4.5 billion. The technology of insulin production has moved ahead and is now produced in human cells besides yeast and other suitable expression systems for increased yield and better efficacy (Walsh 2005). *Arabidopsis thaliana*, lettuce and tobacco plants have been engineered for human insulin production and have been successful (Nykiforuk et al. 2006; Boyhan and Daniell 2011).

3.3 Human Growth Hormone (hGH) by Genentech

The human growth hormone (hGH or somatotropin) is released by the anterior pituitary gland. It is a protein composed of 191 amino acids and triggers cell reproduction, metabolism and overall body growth. Choh Hao Li, a US biochemist of Chinese origin synthesized and purified hGH for the first time (Cole 1996). Thereafter, Genentech pioneered its commercial production using the recombinant DNA approach and sold under the trade name Protropin (1985). The product got discontinued and is now sold as Nutropin since 1993 (Genentech 1993). Besides Genentech, there are other pharmaceutical companies selling recombinant hGH like Pfizer, Roche and NovoNordisk under various trade names. The hormone is administered subcutaneously to children and adults suffering with chronic kidney disease (CKD), Turner's syndrome and growth hormone deficiencies (GHD). The global market sales of hGH are estimated to be generating revenues worth $5261 million by 2026 (TMR 2018).

3.4 Polymerase Chain Reaction (PCR)

Kary Mullis (1983) developed a method for exponential amplification of a piece of DNA using sequence-specific oligonucleotide primer and DNA polymerase (Fig. 5). The process was carried out in a specialized machine (thermal cycler) capable of rapid temperature switching (called thermal cycling). The method is now indispensably used in every basic and medical research laboratory for specific amplification or cloning of genes or any nucleotide sequence. The method proved to be a path-breaking invention in the field of biotechnology which made exact identification of any bacteria, virus, cell type or any gene possible. PCR had an immense impact on state-of-the-art methodologies of molecular biology practised worldwide. Mullis was awarded Nobel Prize in Chemistry in 1993. The PCR technique had direct application in the field of gene cloning, site-directed mutagenesis, parentage identification, prenatal sex determination, pathogen identification, DNA matching from crime scene sample (e.g. blood, semen, hair, skin, etc.), tumour cell testing, etc. For many infectious diseases, cancer stage identification or legal dispute of biological parentage, PCR has now become a standard method of diagnosis and investigation (Bartlett and Stirling 2003).

3.5 The Era of Genomics and Proteomics

Modern times is witnessing tremendous ramification of biotechnology in health care, agriculture and environment sectors. The wealth of data that has been generated by aforementioned discoveries is stored, annotated, analysed and interpreted in various ways. This enormous task has generated an entirely new arm of biotechnology called *bioniformatics*. The *in silico* approach towards solving biotechnology-based questions have been applied in the fields of genetics, molecular biology and protein science. The classical genetics approach aimed at studying one or few "genes to phenotype" on a case to case basis; whereas, *genomics* includes the entire genome characterization in one go. Whole genome sequencing, single nucleotide polymorphism identification, screening of chemical library for fishing out the genomic targets, creation of genomic libraries, etc., all together constitute genomics. The similar approach when applied to study the total RNA population dictated by the genomic

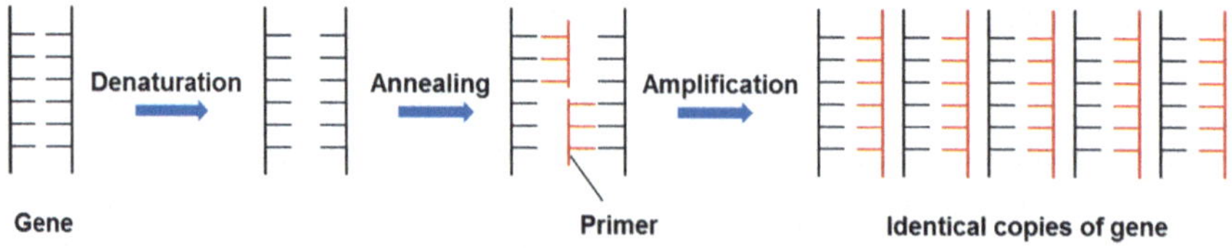

Fig. 5 Diagrammatic representation of the concept of polymerase chain reaction (PCR)

parent is known as *transcriptomics* which includes whole transcriptome profiling, RNA sequencing, microarrays-based gene expression analysis and their validation.

The idea of sequencing the entire human genome was conceived at National Institutes of Health, USA, (1984) from where the original funding for implementation of the idea came. The official launch commenced in 1990 and was termed complete in 2003. A parallel effort was undertaken by a company known as Celera Genomics which was spearheaded by J. Craig Venter. Currently, 92% of human genome stands sequenced which corresponds to euchromatic area; whereas, remaining heterochromatic region still awaits the process. The sequence-related information is stored in the publicly accessible databases like GenBank at NCBI (National Centre for Biotechnology Information, NIH, USA), DDBJ (DNA Database of Japan, National Institutes of Genetics, Shizuoka, Japan) and EMBL (European Molecular Biology Laboratory, Heidelberg, Germany). The three databases frequently exchange information with each other. The information generated by the human genome sequencing was hailed as a historic milestone in the field of biology and medicine. It was envisaged as a tool to identify many oncogenes, developmental disorders-related genes, identification of novel drug targets, throw light on understanding bacterial and viral pathogenesis, etc. (Collins et al. 2007). The information contained in the genome is eventually "executed" by the *proteome* (the sum total of all the proteins present in the cell). The science aimed at studying and a characterization of the proteome is called as proteomics. The term has virtually replaced the phrase protein science. Nowadays, any attempt either to study a single or complex of proteins is included in the term proteomics. Present-day state-of-the-art proteomic methodologies use protein purification strategies, 2-D gel electrophoresis, mass spectrometry analysis, peptide microarrays, Fourier resonance energy transfer (FRET), etc., to identify and characterize the function of proteins (Anderson and Anderson 1998). The protein-related information has been stored in free databases like UniProt Consortium comprising EBI (European Bioinformatics Institute, Hinxton, UK), SIB (Swiss Institute of Bioinformatics, Lausanne, Switzerland) and PIR (Protein Information Resources, Georgetown University, Washington, US) (Wu et al. 2002).

3.6 Metabolomics

The sum total of all the metabolites produced in the cell constitutes the metabolome of the cell. The science involving the study of the metabolome profile of cell or individual is known as metabolomics. The subject matter differs from genomics and proteomics in a way that it studies the actual processes that proteins (which are in turn dictated by the genome) are doing as end products. In this way, it rises above the molecules and takes the investigation to the "systems biology" level. Metabolome directly tells the fate of the cellular processes working in the cell. The study stems from background literature where clinicians or investigators used to analyse the presence of certain compounds in body fluids by the means available in those times. For example, physicians in ancient China employed ants to check the presence of glucose in diabetes patients. Nowadays, modern techniques like gas chromatography–mass spectrometry (GC–MS) and nuclear magnetic resonance (NMR) are used for identification of secondary metabolites in the biological samples (Sussulini 2017). METLIN (2005) was the first database of human metabolome that is maintained by the Scripps Research Institute (La Jolla, USA). It contains the mass spectrometry data (like molecular mass, structure and formulae) of the metabolites (Smith et al. 2005). A human metabolome repository is maintained at University of Alberta, Canada, named Human Metabolome Database (HMDB) that contains information of human body metabolites derived from NMR and GC/LC-MS analyses. At present, it is the most comprehensive and updated metabolomics database that is widely accessed (Wishart et al. 2007).

4 Areas/Branches of Biotechnology

Biotechnology has been classified in multiple ways. Recently, there was a colour-based rainbow-coding of various arms of biotechnology to provide a more holistic view. These colours comprised red, green, blue and white. Each colour specified the particular areas of application of biotechnology under its ambit (Fig. 6), e.g. red was

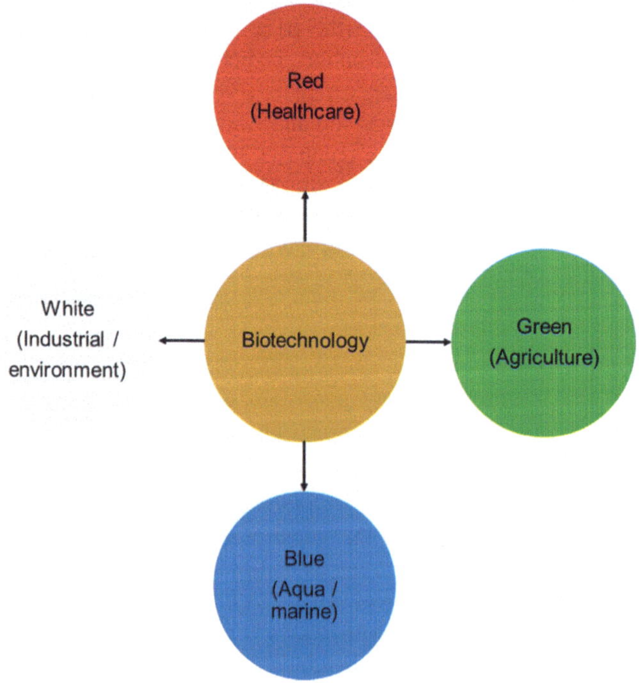

Fig. 6 Schematic description of the branches of biotechnology

constituted by biotech-based biomedical research; agriculture-oriented application was included in green; biotechnological application in marine or aquatic life forms and their usage comprised the blue and that pertaining to industry and environment was termed as white biotechnology (Frazzetto 2003; DaSilva 2004; Kafarski 2012). This section briefly discusses some of the recently emerging representative examples from such categories and highlights their significance in the improvement of lives on earth and its environment.

4.1 Red Biotechnology

The biotechnological inventions and discoveries pertaining to biomedical research or animal health have been collectively included in the umbrella term red biotechnology. It comprises significant developments in the field of stem cell biology, embryo manipulation, transgenic animals (with better nutritive or commercial value; like high milk-yielding cattle), biopharmaceuticals (recombinant vaccines, hormones and other therapeutic proteins production), forensics (DNA fingerprinting), genetic interventions like gene editing and gene therapy, disease diagnostics (pathogen identification, oncogene examination) and personalized medicine (SNPs-related disease predictions in cancers, infection and metabolic disorders).

4.1.1 Genetic Techniques

- **Recombinant DNA technology for production of vaccines**

"*Prevention is better than cure*"—Desiderius Erasmus

Perhaps, the aforementioned statement best describes the significance of vaccines in saving lives from diseases. Ever since Jenner laid the foundation of the field, vaccinology has come a long way from use of live-attenuated or heat-killed vaccines, to recombinantly expressed or the plasmid-based ones. In classical times, infectious virus was passaged for several generations in cell culture to get it mutated, and upon introducing back into the original host it was unable to infect yet elicit an immune response. In this way, it was said to have become "attenuated". Examples of live-attenuated vaccines include polio, rotavirus, measles and H1N1 flu vaccines. Usage of chemicals like formaldehyde or heat to reduce their virulence resulted in inactivated or killed vaccines, e.g. cholera, typhoid or influenza vaccines (Benn et al. 2013).

The recombinant approach applied to manufacture vaccines is twofold: one is called the vectored approach whereby antigenic regions of the pathogen are delivered to patients via bacterial or viral vectors; second approach involves heterologous expression of antigen encoding genes in yeast systems like *Saccharomyces cerevisiae* or *Pichia pastoris* or mammalian and insect cell lines and large-scale purification of the recombinant protein thereafter. The first approach is easy to handle and employs bacterial or viral vector for expressing genes encoding antigenic regions obtained from a wide variety of pathogens. Intramuscularly injected antigens use several viral or bacterial vectors like adenovirus (Ad5 serotype), adeno-associated viruses, recombinant *Mycobacterium bovis* BCG (rBCG strain), *Salmonella* (Shata et al. 2000; Rollier et al. 2011) (Fig. 7b). For antigens requiring post-translational modifications, the first approach proves to be inadequate and therefore use of yeast or mammalian and insect cell lines becomes indispensable. For this, the antigen encoding genes (e.g. Hepatitis B surface antigen) is cloned and expressed in yeast (e.g. *S. cerevisiae*) and cultured in fermenters on a large scale. The recombinant protein assembles in virus-like particles (called VLPs) which are secreted by yeast in the external medium from where it is purified (Fig. 7a). Glaxo Smithkline Plc., London, manufactures hepatitis B vaccine via this approach which is directed against its surface antigen under the brand name Engerix-B (Keating and Noble 2003).

A third category of recombinantly administered vaccines comprises DNA vaccines. The antigen encoding region is expressed in a vector that contains a bacterial origin of replication, a strong viral promoter (e.g. cytomegalovirus, CMV), an MCS region for cloning of the segment and an antibiotic selection marker. Modes of delivery of such vaccines have been intramuscular injection or via gene guns (Fig. 7c). The DNA is adsorbed on gold particles and bombarded on the localized site of action. The idea is to transfer the construct directly in the cells where it expresses and mimics the condition of natural infection. The method has been used for vaccination against TB, leishmaniasis, influenza, HIV, etc. (Yang et al. 1990; Oliveira et al. 1999). Improved methods now incorporate changes made in the construct so as to prevent its degradation inside the cells upon entry or to simultaneously co-express inflammatory cytokines to heighten the immune response (Belakova et al. 2007).

- **Gene editing: CRISPR-Cas** (clustered regularly interspaced short palindromic repeats—CRISPR associated)

Originally discovered as an immune system of bacteria conferring resistance against the invading viral DNA, the technology of CRISPR has been adapted into a toolkit for genome editing. The acronym describes a particular stretch of repetitive nucleotide sequences in the bacterial DNA that became part of the genome because of a previous viral infection. The sequences are capable of degrading viral DNA in case of a future infection by the same or similar DNA virus thus conferring immunity to the bacteria against viral attack. The repeat sequences were discovered in 1987 in *E. coli* by group of investigators in Osaka University, Japan. The group noticed the unusual feature of the repeats which was the presence of interrupting nucleotides between them; however, they could not assign any function to them (Ishino et al. 1987). The acronym CRISPR was given in 2002 by a group that published bacterial genomic loci which harboured interspaced repeat sequences (Jansen et al. 2002). Thereafter, a surge of reports describing the phenomenon was observed in the scientific community which not only established it as a novel form of natural bacterial defence mechanism but also opened the possibility of applying the method for artificial genome editing. The

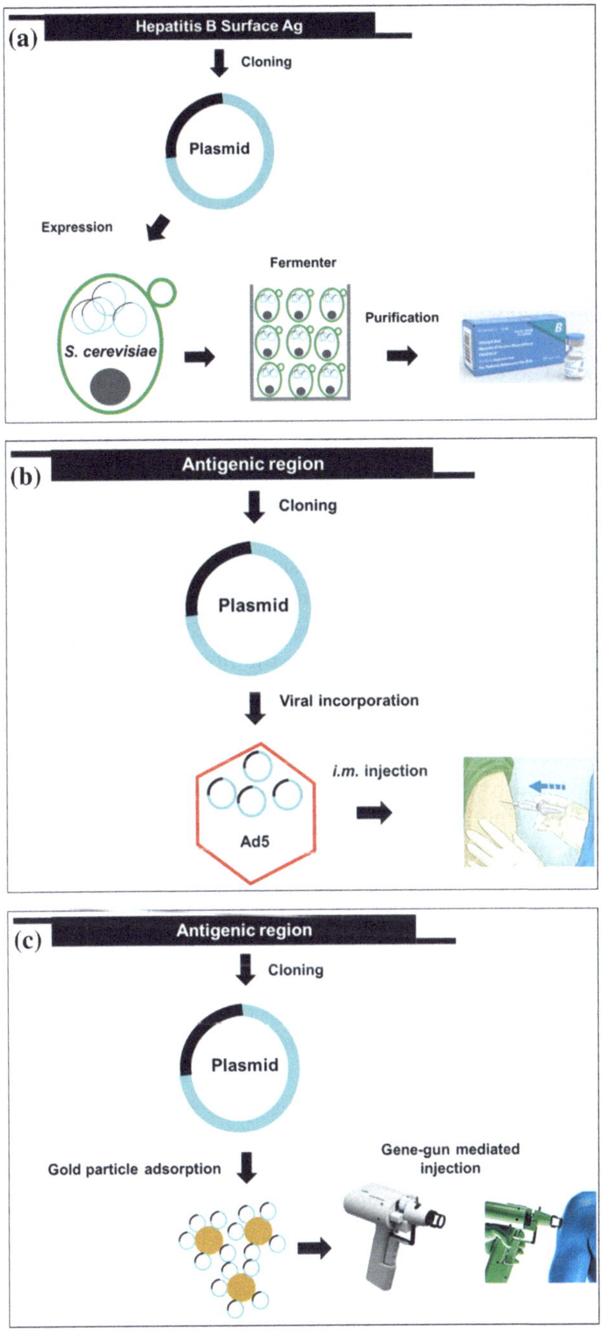

Fig. 7 Schematic describing types of recombinant vaccines. **a** An example of recombinant HbSAg vaccine. Recombinant vaccine administration. **b** Adenovirus (Ad5) mediated recombinant vaccine delivery. **c** Gene gun driven

technology holds immense possibility to correct genetic mutations causing diseases, genetically engineer crops or livestock of desired traits in a cheap and precise manner.

Simply speaking, whenever a DNA virus attacks a bacteria, then bacterial Cas enzyme captures spacer sequences from the invading DNA and integrates into the CRISPR loci in the bacterial genome in arrays. This array encodes an RNA transcript which matures through different types of CRISPR pathways (I, II and III). In type I pathway, the premature or precursor form of CRISPR-RNA or crRNA is cleaved by CRISPR-associated RNAses leading to multiple crRNAs. The type III system employs an unknown RNAse to generated final mature crRNA. The type II system which has been adapted for artificial genome editing (with some modifications) utilizes another trans-activating RNA or tracrRNA which hybridizes with the crRNA and together with Cas endonucleases binds and cuts the target DNA by introducing DSBs (Hsu et al. 2014). The three components, namely crRNA, tracrRNA and Cas9 endonuclease together constitute what is called as CRISPR system which is capable of performing its function even in vitro (Gasiunas et al. 2012; Jinek et al. 2012). For artificial purposes, the former two components have been combined, and a single guide RNA or sgRNA has been used successfully (Jinek et al. 2012). These findings have paved the way of translating the basic knowledge of this phenomenon into useful biomedical and agricultural applications.

In health care, the method has been applied to correct the genetic mutations causing Duchenne's muscular dystrophy (DMD), cystic fibrosis, haematological malignancies and HIV pathogenesis. Using the CRISPR technology in mouse model system, DMD gene expression was rescued by delivering its functional copy through adeno-associated virus vector. The muscular abilities of mice were found to be partially restored as compared to the control group exhibiting degenerating muscle conditions (Tabebordbar et al. 2016). The genome of HIV has been targeted to stop its replication in the infected cells without any toxicity. The achievement could be viewed as a significant advance over the currently practised method to contain or stop viral replication in preclinical systems (Hu et al. 2014). In a parallel study, the CXCR4 receptor of HIV was removed successfully from human T cells using the CRISPR system; a feat which could be mobilized vertically for edited bone marrow transplants from patients suffering from AIDS (Schumann et al. 2015).

There is growing concern about the potential misuse of CRISPR-Cas technology, particularly for the purpose of editing the genome of human embryos. Human embryo manipulation is legal in China and several states of USA. However, there are clear ethical concerns regarding the reckless use of the technology towards giving birth to only the ones with modified features. Such changes when introduced are heritable and are passed on for generations with unpredictable future outcomes. There has been a case of human embryo editing which attracted worldwide shock and condemnation. It called for an urgent action against the erring group simultaneous with rounds of brainstorm sessions held globally to chart out regulation(s) to prevent misuse of CRISPR technology (Cyranoski and Reardon 2015; Liang et al. 2015). Nevertheless, the method has invited global interest towards harnessing its potential in drug development by pharmaceutical companies, improvement of crop traits by agriculture-based industries and general biomedical research towards understanding of biological processes governing several less characterized disease conditions.

- **Gene therapy**

The introduction of DNA into cells or organisms for therapeutic benefits is known as gene therapy. The first idea of gene therapy was proposed by Theodore Friedmann and Richard Roblin in 1972 by emphasizing on the beneficial aspects of supplying exogenous DNA to the organism for correcting the genetic defects (Friedmann and Roblin 1972). There are two types of gene therapy strategies, namely somatic cell gene therapy and germline gene therapy. Transfer of therapeutic DNA/gene into any cell other than germ cells or progenitor stem cells is called as somatic cell gene therapy. Such interventions do not affect the germ cells of the person and as such results are not heritable (Mavilio and Ferrari 2008). Two modes of delivery of the therapeutic DNA exists, namely in vivo and ex vivo. The delivery of the DNA directly into the patient's or model system's body is known as in vivo approach; whereas, transfer of gene into cells isolated from the body and then introducing them back after ascertaining its stable expression is called as ex vivo mode of gene therapy. The method of delivery of the DNA is divided into two categories: viral-dependent vectors and viral-independent vectors. The former employs the use of viruses devoid of their original genome and instead inserted with the functional copy of the gene for its delivery either into the body directly or in cultured cells. The latter group comprises liposome, gene gun, electroporation or dendrimers-based DNA transfer into the organism or cultured cells. Examples of viruses utilized for gene therapy include herpes simplex, human immunodeficiency, adeno-associated and vaccinia viruses (Mavilio and Ferrari 2008).

Adenosine deaminase (ADA) deficiency caused severe combined immunodeficiency (SCID); an autosomal recessive disorder (Hirschhorn et al. 1979) was the first target of somatic gene therapy in a clinical trial performed by R. Michael Blaese, W. French Anderson and Kenneth Culver in 1990. It was characterized by the lack of ADA enzyme which was essential for DNA synthesis. Using viral vector, the gene encoding the enzyme was delivered to the patient (Culver et al. 1990; Rosenberg et al. 1990). Haematological disorder like beta-thalassaemia is caused by reduced or no production of beta globin chain of haemoglobin. Using lentiviral vector-mediated beta globin gene delivery led to successful production of correct haemoglobin (Sadelain 2006). Gendicine was the first gene therapy drug approved for the treatment of head and neck squamous cell carcinoma in China. The drug enters the cells via receptor-mediated endocytosis and overexpresses p53 levels (Pearson et al. 2004).

Despite endowed with enormous potential of treating diseases and disorders, gene therapy has been marred by several challenges that has led to its decline in recent years. The lack of persistence of effect is the first of all issues that has been slowing down the pace of gene therapy. The functional copy of the administered gene is lost with successive cell divisions which lead to reducing effect of the gene therapy. As a result, the patient needs to get administered the therapy multiple times so as to maintain the therapeutic effect. Secondly, problem of immune challenge caused by viral vectors poses a serious impediment towards the usage of such vehicles for gene delivery. The surface composition and pattern of many viruses elicit major immune response in the person administered with viral vector-based gene therapy. A second approach of using non-viral vectors has not met with much success. The low rates of gene delivery through such approaches are a major obstacle which has prevented their large-scale testing (Goncalves and Paiva 2017).

- **DNA fingerprinting or profiling**

The use of unique DNA sequence properties to identify an individual is known DNA profiling or fingerprinting. The method was originally discovered by Sir Prof. Alec Jefferys in 1984 at the University of Leicester, UK. The genome between two unrelated individuals match up to 99% yet the unmatched portion offers enough to discriminate between them. Professor Jefferys also observed that the fingerprint in a child comprised half from both the parents. The classical method of fingerprinting relies on the selective detection of certain unique repetitive DNA sequences called as variable number tandem repeats (VNTRs) which are present in the genome in the form of minisatellites and microsatellites (Roewer 2013). The technique involved extraction of DNA from samples (e.g. blood, hair follicle, semen, nails or other body parts or secretions) and its restriction digestion and electrophoresis to generate band fragments on the basis of restriction sites present. The sample was then blotted to nitrocellulose membrane and repetitive sequences were detected using specific complementary radiolabelled probes. The minisatellites are six to hundred nucleotides long and repeat for hundreds of times in the genome (Tautz 1993). Microsatellites (also known as short tandem repeats or STRs) on the other hand are shorter in length (as the name suggests) ranging from one to five nucleotides repeating for some hundred times (Koreth et al. 1996).

The aforementioned method has now been replaced by a PCR-based technique which involves DNA extraction followed by amplification of microsatellite region. This has greatly improved the workflow of fingerprinting and has added to its rise in success. The method now is widely used for parentage identification, cell lines authentication, to check cancer progression but perhaps its greatest application has been in the field of forensics where it is used for identification of criminal. Owing to this, now many countries maintain a DNA database of their population with which the fingerprinting results are compared to find out the real culprit. In the USA, Coding for DNA Identification System (CODIS) is maintained by Federal Bureau of Investigation (FBI) for the identification of criminals by using DNA fingerprinting (Saad 2005).

The method also revealed through arbitrarily primed-PCR (AP-PCR), the mutations in the microsatellite regions of cancer tissues and established the process of a new mechanism of carcinogenesis. Since then fingerprinting has been successfully used for identification of cancer mutations in the microsatellites thereby allowing to identify the stages of progression (Perucho 1996). In an earlier study, 46 different cell lines (including cancer cells) were authenticated by using DNA fingerprinting of minisatellites (Gilbert et al. 1990). The case of disputed paternity is now routinely solved based on DNA profiling as mandated by the courts worldwide. As mentioned earlier, based on the shared fingerprint of the VNTRs in the child and the suspected person, it is possible to nail down the biological father.

4.1.2 Embryological/Organismal Manipulation

- **Somatic cell nuclear transfer (SCNT)**

The technique involves transfer of a nucleus from donor cell to enucleated egg cell or ovum and allowing it to develop into an embryo. After many rounds of mitotic divisions, the embryo reaches blastula stage where it develops inner cell mass (ICM) containing embryonic stem cells (ESCs). From here, as the need demands the embryo can be implanted into a female animal if reproductive cloning is the objective or for therapeutic cloning, the ESCs can be extracted out and can be used for tissue regeneration by harnessing their pluripotent nature (Fig. 8) (McLaren 2000). Sir Hans Spemann (1928), a German embryologist, is credited with the discovery of the then *embryonic induction*, an idea considered to be the predecessor of modern-day SCNT. Through microsurgical needle, Spemann and his colleagues inserted a particular region of an embryo into another embryo which led to development of a new embryonic growth irrespective of the area where it was transplanted (Spemann 1938). Dolly (sheep) was the first organism born out of reproductive cloning through SCNT method in 1997 at Roslin Institute, The University of Edinburgh, UK. The investigators incorporated a nucleus from an adult cell into an enucleated ovum and allowed to develop the zygote till blastula stage of embryogenesis. Later, it was implanted into a female sheep that served as a surrogate mother for the developing foetus and gave birth to Dolly (Edwards 1999). Very recently, macaque monkeys

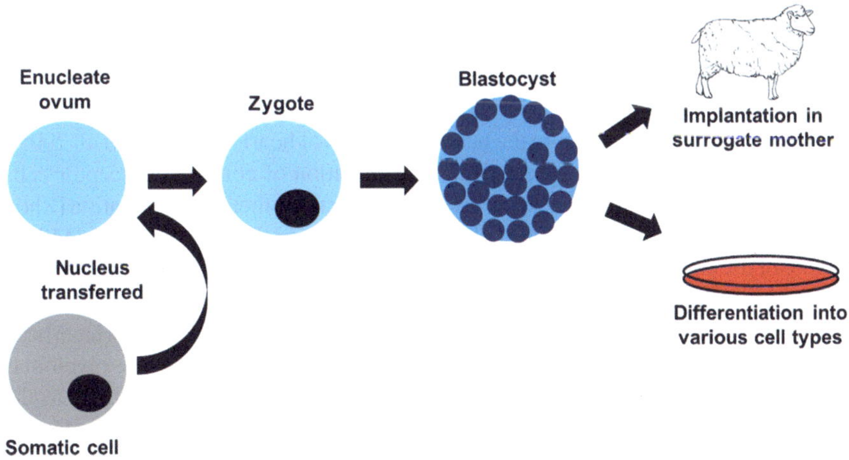

Fig. 8 Schematic depicting somatic cell nuclear transfer

were cloned in China which marked the first successful reproductive cloning using SCNT in primate species (Liu et al. 2018).

Therapeutic cloning is perhaps the best legacy that SCNT has left behind for the welfare of mankind in general. The power of generating blastocyst and extraction of ICM allowed the investigators to generate multiple tissues and organs of human body which has direct application in the treatment of several diseases. Researchers have been successful in generating pancreatic endocrine cells by the application of SCNT technique which were found to be capable of producing hormones like insulin, glucagon and ghrelin. An advancement like this holds immense possibility towards better treatment of type 1 diabetes in which insulin secretion is compromised (D'Amour et al. 2006). In another study done in mouse, investigators were able to regenerate spinal cord using motor neurons from ESCs, an achievement of medical importance in cases of paralysis (Liang et al. 2006).

- **Stem cells and their applications**

Stem cells are capable of differentiating into almost any cell type of the body and can also regenerate themselves. The property by which a stem cell can differentiate into myriad cell types is known as pluripotency, and such cells are known as pluripotent. As mentioned in the previous section, the inner cell mass (ICM) of a blastocyst houses the embryonic stem cells (ESCs) capable of generating all three germ layers and their cell types (Thomson et al. 1998). Besides the ESCs, there is another method of generating stem cells which involves reprogramming back a differentiated cell into a state of stemness. The stem cells thus created are known as induced pluripotent stem cell (Aoi et al. 2008; Chagastelles and Nardi 2011). Apart from stem cells derived from the embryo, the bone marrow also contains a population of haematopoietic stem cells (HSCs) which give rise to all the blood cell types. HSCs have been clinically used since 1960s (Good et al. 1969) and are now obtained from umbilical cord and placenta (Kogler et al. 2004).

The clinical uses of stem cells have opened a new avenue of research which has great potential for offering cure against several diseases. Cardiomyocytes generated from human ESCs (hESCs) have been successfully transplanted into mouse model system and were found to restore the beating function of heart (Laflamme et al. 2005). hESCs incorporated with gene essential for generation of cone photoreceptor cells have been successfully used for treating retinal pigment epithelial degeneration (Zhou et al. 2015). Another class of stem cells, known as mesenchymal stem cells (MSCs; capable of differentiating into cells of mesodermal origin) derived from bone marrow has been used to generate bladder tissue in baboons thus establishing a non-human primate model for potential human applications (Sharma et al. 2011). HSCs have been used for transplantation in patients suffering from various lymphomas, myelomas and immunodeficiency diseases (Eaves 2015). Prior to administering HSCs to the patient, the bone marrow is destroyed by high doses of chemotherapeutic drugs with or without radiotherapy (myeloablation). Thereafter, HSCs obtained from the patient are injected in the bloodstream from where they are able to reach and replenish the destroyed tissue. This mode of administering patient's own HSCs to him or her is known as autologous as compared to allogenic in which HSCs are obtained from healthy donor with HLA match type (Russell et al. 2000). As pointed out earlier, mother's umbilical cord and placenta are two prime organs which are storehouses of HSCs; therefore, people are now getting their cord blood cells stored in blood banks. The stem cells derived from such banks can be very helpful in case of future haematological malignancies, inherited disorders or other genetic diseases.

4.1.3 Molecular Diagnostics

The use of DNA/RNA or protein(s) for medical diagnosis or prognosis of diseases or disorders is collectively referred as molecular diagnostics. The methods commonly utilize the detection of DNA or RNA converted to cDNA through PCR or hybridization-based techniques. Proteins are generally detected by using antibodies and antibody-dependent techniques like enzyme-linked immunosorbent assay (ELISA) or flow cytometry in a method called as immunophenotyping (Poste 2001). The methods are fast, precise and specific as compared to traditional diagnostic approaches.

- **PCR-based detection**

Since the inception of PCR by Kary Mullis, the method has evolved into a more advanced technique which has enabled the detection of DNA or RNA in real time through a technique called Real-Time or Quantitative PCR (qPCR). qPCR works on the principle of usual PCR except that the amplified product can be visualized by using fluorescent dye (e.g. SYBR green) or fluorescent-labelled complementary probes (TaqMan probes) whose signal rises as the amplified product accumulates (Kubista et al. 2006). qPCR is of great relevance in the field of molecular and diagnostic virology where one can identify viral serotypes in the patient's body sample. Since clinical samples are often limited, a PCR-based method of detection proves to be of immense advantage because the target gene can be amplified exponentially and can be monitored live. The method has been successfully applied to detect and quantify causative agents of several infectious diseases; viruses like HIV, hepatitis B virus (HBV), herpes simplex virus (HSV) 1 and 2, human papilloma virus (HPV) in patient's samples (Valones et al. 2009). For HIV, p24 gene is routinely used for amplification for detecting the presence of HIV-1 in patient's dried blood spots (Lakshmi et al. 2011). All four serotypes of dengue virus (DENV 1–4) are routinely detected in clinical laboratory using PCR methodology to amplify C and E genes using serotype-specific primers (Das et al. 2008). HBV has been successfully detected using TaqMan probe-based qPCR technique in clinical samples (Zhao et al. 2005). For bacterial strain identification, 16S rRNA gene (considered to be universal reference) of each strain is amplified through PCR. Bacteria like *Mycobacterium tuberculosis* and *Helicobacter pylori* are detected by their sequence-specific primers against 16S rRNA (Barghouthi 2011). Another target is the outer membrane protein (OMP) gene routinely used for detection of bacteria belonging to *Chlamydia* sp. which primarily causes respiratory illnesses. Many species of Chlamydia have been known to cause Alzheimer's, arthritis, atherosclerosis, etc. (Yamamoto 2002).

- **FISH (fluorescence in situ hybridization)**

The technique of FISH involves hybridization of fluorescently conjugated DNA/RNA probes with its target genomic region or mRNA. Its analysis requires an imaging system like a fluorescence microscope. The method can be performed on cells and fixed tissues. The samples (cells or fixed tissues) are permeabilized and incubated with the fluorescently conjugated DNA or RNA probes which then hybridize with their target. FISH-based diagnostic tests are routinely used for detection of HER2 expression levels in various cancers (Hicks and Tubbs 2005). FISH probes are widely used for detecting chromosomal translocations in the case of lymphomas (Tempest et al. 2008). Using the probes, BCR-ABL translocation gene has been successfully detected in peripheral blood granulocytes of patients suffering from chronic myelogenous leukaemia (CML) (Takahashi et al. 2005).

- **Immunophenotyping**

The method of detection of surface proteins through antibodies and flow cytometry is known as immunophenotyping. The method involves labelling cells with their respective cell surface markers, e.g. CD markers; CD8 in case of T cells or any other cell surface-expressed protein used for identification of cell type and developmental stage. Both high and low expression profiles of the markers are used as indicator of particular cell type and its developmental stage. These two parameters help in the diagnosis and prognosis of the disorder, respectively. The technique allows to discriminate between various stages of cells based on their development stage specific surface marker profile. This is helpful in deciding the prognostic strategy when a patient is undergoing therapy. The technique also enables to identify the lineage between cells of B or T lymphocyte origin. The method has been of great use in haematological disorders including malignancies like chronic lymphocytic leukaemia (CLL) (Ivancevic et al. 2014). Cancer cells tend to acquire drug resistance over the period of therapy which can be ascertained by immunophenotyping P-glycoprotein (MDR-1) or multidrug resistance-associated protein (MRP-1) (Jakab et al. 2005). In the case of HIV infection, $CD8^+$ T lymphocytes decrease in population which can be accurately diagnosed by flow cytometry-based immunophenotyping. B cell lymphomas are detected by low expression profile of CD20 marker. Using the method, leucocyte adhesion deficiency disorders can be identified accurately by looking at the decreased expression profile of the leucocyte β2 integrin receptor complex which comprises CD11/CD18 complex (Brown and Wittwer 2000).

4.1.4 Personalized Medicine

The utilization and practice of the knowledge of genetic information of a patient to evaluate disease risk assessment, drug response and deciding the therapy are known as personalized medicine (Mancinelli et al. 2000). Through genetic analysis like whole genome sequencing, it is possible to identify variations in the genes encoding receptors or enzymes that may predict the outcome of a particular drug treatment.

Advancement in the field of genomics after the completion of Human Genome Project has led to the generation of vast knowledge regarding underlying genetic differences in a particular population. This knowledge when applied to study disease-related predisposition markers or drug response of patients is known as pharmacogenomics (Vogenberg et al. 2010).

- **Genome-wide association studies (GWAS)**

The comparison of genetic differences associated with any particular trait or disease condition in a population between healthy and diseased subjects is known as genome-wide association studies. The common genetic differences taken into account while doing these analyses are single nucleotide polymorphisms (SNPs) which have been shown to be present in varying frequencies in a population of healthy and diseased individuals. Almost one million SNPs can be genotyped in one scan of a patient's DNA sample (Spencer et al. 2009). The first GWA study was reported in 2005 in patients suffering from age-related macular degeneration (leads to blindness in elder people). The study identified two SNPs present in complement factor H gene which were linked with the increased risk of the disease (Haines et al. 2005).

In a separate study performed by the Wellcome Trust Case Control Consortium (as reported in 2007), around 2000 patients for each of seven diseases, namely type 1 diabetes, type 2 diabetes, Crohn's disease, rheumatoid arthritis, bipolar disorder, coroner artery disease and hypertension were analysed for the variation in their SNPs using Affymetrix SNP array. This comprehensive effort revealed several new genetic loci associated with risk of diseases under study (Consortium 2007).

4.2 Green Biotechnology

Application of biotechnology in agriculture sector with the aim of increasing the productivity and ensuring food security is called as green biotechnology (Kafarski 2012). The methodology adopted for realization of aforesaid goal is somewhat similar to what is used in healthcare sector or red biotechnology due to the universality of molecular biology principles. Broadly, the branch includes gene manipulations done for crop or livestock improvement by increasing the crop or animal-based yield or conferring resistance against biotic or abiotic stress factors.

4.2.1 Genetic Transformation in Plants

There are a number of methods for incorporating foreign genes in plant systems, namely *Agrobacterium tumefaciens*-mediated gene transfer, electroporation, gene gun or biolistics and microinjection (Lorence and Verpoorte 2004). Among all these, *Agrobacterium*-dependent DNA transfer has been the most successful method, especially in dicotyledonous plants. The bacterium is a natural plant pathogen that upon infection transfers a portion of tumour-inducing plasmid (Ti plasmid) DNA called

T-DNA, which then triggers a tumour-like growth (called crown gall disease) in the plant (Nester Gordon et al. 1984). The T-DNA is flanked by border repeat sequences, which are retained while replacing the rest of T-DNA with the gene of interest for subsequent infection and incorporation of the gene by *A. tumefaciens* in the cultured plant cells (Quispe-Huamanquispe et al. 2017). Other technique like gene gun-dependent DNA transfer involves coating gold or tungsten particle with the desired plasmid DNA. The particles are then shot in the growing cultured cells for incorporation in the genome. *Bt* maize is an example of transgenic plant derived from gene gun-mediated DNA transfer (Slater et al. 2008).

- **Transgenic plants**

Also called genetically modified or engineered crops (GM/GE crops) are created by transferring genes to organisms from a different species which confers a trait not found in the recipient's species. The transgenic plants have been successfully created and field tried for their property to yield more, resist herbicides or insects, be resistant to drought or high soil salinity conditions, etc. (Banjara et al. 2012). The total produce of GM crops has increased from 1.7 million in 1996 to more than 175 million by 2013. The USA leads in the use of biotech crops (cotton, soybean, maize and canola) by accounting for over 40% of global produce. India and China are the biggest growers of Bt cotton (Clive 2015). *Gossypium* (cotton) species was genetically altered to express Cry protein (endotoxin) of *Bacillus thuringiensis* which is highly effective in killing lepidopteran insects. The crop was first commercially introduced by Monsanto, Inc. in the USA, in 1996 and later in India in 2003 (Rocha-Munive et al. 2018).

Crop plants of nutritional importance like rice have been genetically modified to produce golden rice variety. *Oryza sativa* (rice) was transformed into biosynthetic gene encoding β carotene which led to its higher yield as compared to wild-type strain. Two varieties were generated of different yield values of β carotene: golden rice-1 yielding 1.6 μg of vitamin A per gram of rice by transferring genes from daffodil and golden rice-2 yielding 35 μg of vitamin A per gram of rice by transferring genes from maize. The varieties are of considerable importance towards fulfilment of vitamin A deficiency which causes night blindness and xerophthalmia (Tang et al. 2009).

Transgenic plants of ornamental importance are now being generated with increased floral scent content or changed colours of flowers. The floral scents in the flowers are due to the presence of volatile organic compounds, namely terpenoids, benzenoid and aromatic amino acids (Piechulla and Effmert 2010). *BEAT* gene has successfully transferred from *Clarkia breweri* into *Eustoma grandiflorum* which induced fragrance in the petals of the recipient plant (Aranovich et al. 2007).

4.2.2 Genetic Modifications in Livestock

Microinjecting the embryos with foreign DNA or somatic cell nuclear transfer is the most feasible methods for developing transgenic animals of agricultural importance. The first method being the most widely practised in which embryos are microinjected with the gene of interest cloned suitably in an expression cassette and implanted back in the female (Kues and Niemann 2011). The progeny is screened for the presence of the transgene by southern hybridization, pcr or DNA sequencing (Setlow and Hollaender 2002). The main objectives behind such modifications are increased milk and meat production, to render the livestock disease-free and utilize the animal for production of therapeutic proteins through the process called *biopharming*.

- **Recombinant human lactoferrin**

Human lactoferrin (hLF) is an 80 kD iron-binding glycoprotein present in milk and multiple bodily secretions. It is essential for iron uptake in body and kills several harmful iron requiring bacteria by sequestrating iron away from them (Lonnerdal and Iyer 1995). Using native gene (comprising all introns and exons) instead of only cDNAs encoding the protein has been found to be yielding higher amounts of therapeutic proteins in the milk of mammals (Choi et al. 1991; Whitelaw et al. 1991). With the former approach, whole genes (having introns and exons) encoding human lactoferrin, transgenic cows have been produced, via both microinjection and nuclear transfer methods yielding ~3–3.4 mg/ml of human lactoferrin in their milk (van Berkel et al. 2002; Yang et al. 2008). The latter study utilized bacterial artificial chromosome for accommodating the hLF construct and bovine fibroblast cells for microinjecting the DNA. Somatic cell nuclear transfer led to the birth of two transgenic cows secreting levels of hLF as mentioned earlier. Biochemical assays revealed their iron-binding and release efficiency similar to the natural hLF.

- **Recombinant antithrombin**

In 2006, the European Medicines Agency and later in 2009, US FDA approved the first transgenically obtained drug—a recombinant human antithrombin III protein derived from the milk of goat, for surgical and childbirth applications. The transgenic goat was developed by the then GTC Biotherapeutics, Inc. at Massachusetts (USA) in collaboration with Louisiana State University Agriculture Center (USA) by injecting goat cells with human antithrombin gene (RD 2004). The protein is purified from goat's milk and is sold currently under the brand name *ATryn* by rEVO Biologics, Southborough, MA (USA) (Erickson 2009). The drug is a lifesaver for those who suffer from hereditary antithrombin deficiency which puts them at the risk of developing deep vein thrombosis or pulmonary embolism (Maksimenko et al. 2013).

- **Transgenic Salmon fish**

AquAdvantage or transgenic salmon fish was developed at Aqua Bounty in Canada in 2006. Wild-type Atlantic salmon (*Salmo salar*) eggs were injected with the growth hormone encoding gene from Chinook salmon (*Oncorhynchus tshawytscha*) under the regulatory elements of an antifreeze protein derived from an ocean pout (*Zoarces americanus*). Whereas a wild-type salmon feeds only in spring or summer and takes 3 years to attain its full length, the transgenically created salmon could reach the same length in less than two years. These fishes have triploid genomic content as compared to their diploid wild-type counterparts and are sterile, thus preventing the risk of interbreeding (Yaskowiak et al. 2006). The US FDA approved AquAdvantage for human consumption only in 2015 while Canada did the same six months later (Waltz 2017).

4.2.3 Germplasm Conservation

For stable maintenance of the transgenic or wild-type traits of plants and animals, it is essential to preserve the organisms' genetic information so that the traits can pass on from generation to generation without getting exhausted. The body parts containing such genetic information which are viable despite being preserved at ultra-low temperature for long periods of time are collectively called as germplasm. For plants, seeds, calli, pollen, excised embryos or root/shoot bud tips comprise the germplasm. Animal germplasm includes semen, oocytes or embryos. With the advancement of techniques like artificial insemination, the need of storage and preservation of male and female gametes was realized. There are two principal modes of germplasm conservation of both plants and animals, namely in situ and ex situ. The preservation of the germplasm in the natural habitat of the organism is called as in situ; whereas, the artificial methods executed outside the boundary of natural habitat are known as ex situ modes of conservation, respectively. The method of cryopreservation is an example of the latter whereby the hereditary information containing components of plant or animal bodies is mixed with a cryoprotectant chemical and frozen in liquid nitrogen (Ruane et al. 2006).

Cryopreservation is achieved via two routes: slow freezing and vitrification (flash freezing). In the former method, the rapid formation of ice crystal is avoided by using chemicals like polyvinyl alcohol or synthetic biopolymers like alginates. The rate of cooling is brought down to 1 °C per minute using a freezing box and cryoprotectants like glycerol and dimethyl sulfoxide (Vutyavanich et al. 2010; Sambu 2015). Almost majority of animal cells, tissues and embryos are frozen by this method. Vitrification, also known as flash freezing, on the other hand bypasses the formation of ice crystals at all. The sudden plunge in ultra-low temperature (megakelvins per second) in cryoprotectant (e.g. ethylene glycol or sucrose) leads to amorphous ice formation which is different from ice crystal sheet formation in which water molecules arrange to create hexagonal lattice. The method was first used to freeze human oocytes which were used to deliver a healthy baby girl through in vitro fertilization (Kuleshova et al. 1999).

4.3 Blue Biotechnology

Officially, the term blue biotechnology has not been defined yet by any government/non-governmental organization. Although, the realm of blue biotechnology encompasses marine bioresources and their applications, a more widely acceptable definition of marine biotechnology is used alternatively for the purpose. The European Marine Board defines the area of marine biotechnology as following:

> The application of science and technology for the production of knowledge, goods and services from (marine) biological resources. (Adapted from the Organization for Economic Co-operation and Development general definition of biotechnology 2005)

The field is relatively nascent as compared to other areas of biotechnology and not entirely independent from the rest. Blue biotechnology has clear overlaps with other areas of biotechnology because of the technical know-how derived from them. The marine realm of the biosphere comprises 70% of the total earth's resources. Most of the phyla known till date belong to this realm. A majority of them have been used for societal applications. Broadly speaking, marine microorganisms, phytoplanktons, red, green and brown algae, cnidarians, sponges, echinoderms and mangroves are a few classes from which products of commercial value have been extracted (Blunt et al. 2013). Figure 9 describes the phyla wise contribution of marine life forms for biotechnological, industrial or domestic purposes (Arrieta et al. 2010).

The use of marine life forms on such an enormous scale has led to rise in marine organism's gene sequence-related patents being filed/granted at the International Patents Office. These approvals have been granted in different sectors of biotechnology affecting various aspects of human life (Arrieta et al. 2010). Most of these range from the patents impacting human health to the field of biofuel generation as shown in Fig. 10.

The field starts with the discovery and bioprospecting of marine bioresources with potential biotechnological application. The commercial potential of these resources

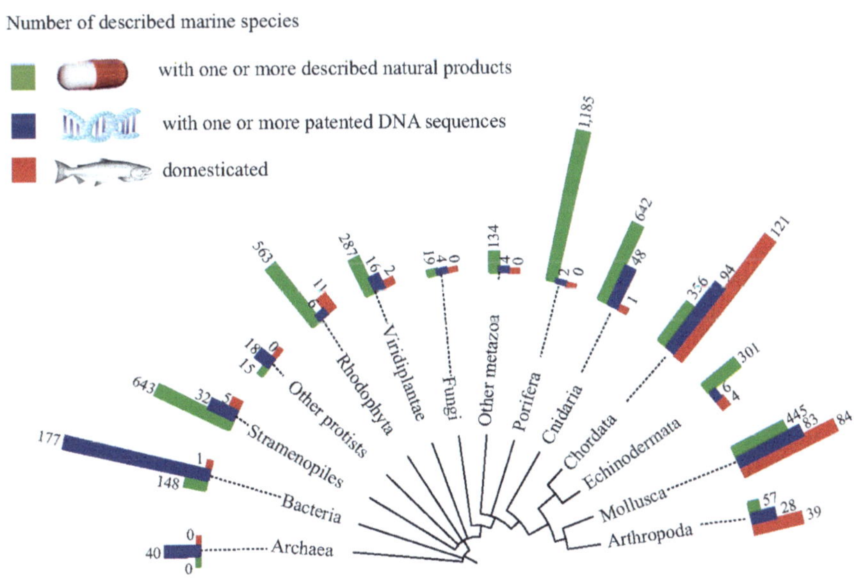

Fig. 9 Schematic depicting marine life forms usage for biotechnology or other human benefits. Each histogram indicates per cent organisms present in the respective taxon utilized for indicated purposes shown in the upper legend (reproduced with kind permissions from the National Academy of Sciences, USA and Dr. Jesus M Arrieta, Spain; (Arrieta et al. 2010)

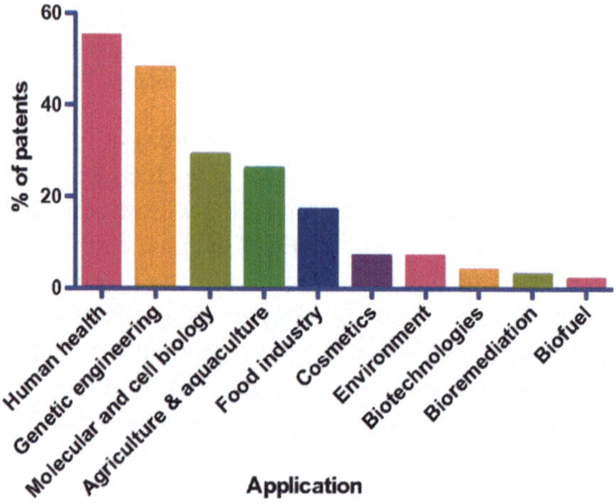

Fig. 10 Graphical representation of distribution of 460 patents deposited with International Patents Office in the area of marine biotechnology. Since one patent may fall in more than one category, sum total percentage exceeds 100% (reproduced with kind permissions from the National Academy of Sciences, USA and Dr. Jesus M Arrieta, Spain; (Arrieta et al. 2010)

ranges from health care in the form of novel pharmacologically active substances to the development of biofuels for providing clean energy. The European biotechnology industry generates annual revenue of € 754 million from the blue biotech sector (ECORYS 2014; ERA-NET 2017). Identification of key gene(s) and or other biologically active molecule and their characterization forms the next step. This involves the screening, selecting and identifying vast number of marine organisms for desired activity. Such organisms range from algal seaweeds to bacteria residing in deep sea hydrothermal vents. Usually, such organisms are difficult to grow in laboratory due to the requirement of pressure, temperature and light conditions in which they otherwise grow. Upon the establishment of successful growth conditions, the organism's DNA is isolated for the purpose of identification of the potential gene encoding the desired product. If the desired product is a biochemical, then its laboratory scale purification is undertaken. The gene or chemical compound identified in such manner is sequenced or assayed for activity in laboratories. Genes are matched to known databases to find out the uniqueness which indicates their evolutionary difference with other known organisms. An example is the famous GFP gene isolated from a bioluminescent marine cnidarian (*Aequorea victoria*), encoding a green fluorescent protein utilized in biomedical research.

After ascertaining the sequence, further research and development programmes are undertaken to characterize the genomic sequence for its expression, purification of the protein and standardization of biological/toxicological assays, and the efficacy in the biological process is formally reported. Examination of commercial feasibility is the deciding factor for the technology adaptation for upscaling to industries and sometimes also the major bottleneck. However, there are several products of marine origin which have seen the light of day, and these are being sold as food, medicine, cosmetic ingredients, biofuels and bioremediation of the polluted environment. Both marine flora and fauna have contributed to the expanding market of such commercial products on which several industries falling in aforementioned sectors thrive. Alginates are extracted from brown seaweeds (class: Phaeophyceae) like *Laminaria, Macrocystis and Ascophyllum* and are used in food, textile and pharmaceutical industries. In food industries, it is used for increasing the viscosity like in making jellies. Another brown sea algae, *Fucus vesiculosus*, is used in its extract form as a dietary supplement. Algae-derived oils are good source of omega-3-fatty acids (monounsaturated and polyunsaturated fatty acids; EPA and DHA) that equates to oils obtained from salmon fish (Doughman et al. 2007). Microalgae, *Haematococcus pluvialis*, yields a red-coloured pigment called astaxanthin that is used in the manufacturing of fish feed and is sold by Fuji Chemical Ltd., Japan. Calcium alginates have been used in skin wound dressings for healing and micro-encapsulation of natural polyphenols from wine wastes (Lansdown 2002; Aizpurua-Olaizola et al. 2016). *Ecteinascidia turbinata*, a sea squirt, was identified by the National Cancer Institute (NIH, USA) for producing an anticancer compound. The compound ecteinascidin 743 was purified many years later and is now commercially sold as Trabectedin by PharmaMar S.A. (Spain) (Rinehart 2000). Ziconotide is a powerful analgesic drug which was first isolated from cone snail (*Conus magus*) in early 1980s by scientist Michael McIntosh at University of Utah (McIntosh et al. 1982). The compound which is a peptide got an FDA approval in 2004 and was sold by the name Prialt by Elan Pharmaceuticals, Ireland (Pope and Deer 2013).

Cosmetic industry has been deriving a considerable amount of raw material from marine bioresources. Brown algae, *Ascophyllum nodosum* and *Halopteris scoparia* (class: Phaeophyceae), contains compounds that protect skin from harmful UV rays of sun. The algal extract mix has been patented by Gelyma (France) and is sold by the name Actiseane by Biosil Technologies Inc., NJ (USA) (Andre et al. 2002). *Alcanivorax borkumensis* is a marine flagellar bacterium whose genome contains sequences encoding petroleum degrading gene products. The bacterium uses alkanes as its diet and therefore is employed for degradation of oil spills in oceans. It is therefore also known as "oil-eating" bacterium (Schneiker et al. 2006). Mutant form of *Alcanivorax* is used for production of polyhydroxy alkanoates (PHA) which are used for manufacturing biodegradable bioplastics (Sabirova et al. 2006). Several marine algae are currently explored and used as replacement of tradition fossil fuels for powering engines. *Botryococcus braunii* (family: Trebouxiophyceae) a colonial green microalga is known to secrete hydrocarbon rich oils in the exterior which has been recently purified using electric currents to disrupt the bacterial colony (Banerjee et al. 2002; Guionet et al. 2017).

4.4 White Biotechnology

The term white biotechnology is essentially European in origin. It is alternatively called as industrial biotechnology which simply means application of living cells, microbes, yeast, fungi, plants and enzymes, etc., for industrial applications (Lee and Jang 2006a, b; Ribeiro et al. 2016). Microbes and enzymes are collectively referred to as *biocatalysts* which are used to manufacture paper, antibiotics, biopolymers, etc. (Heux et al. 2015). Since long time, even before the coinage of the term white biotechnology, living organisms have been used to manufacture products on an industrial scale by techniques like fermentation for making ethanol, acetic acid and various secondary metabolites. Fermentation simply means the decomposition of complex organic compounds to simpler products by enzymes secreted by bacteria and yeast. Modern-day reaction takes place in a steel vessel called the fermenter. The medium for fermentation is composed of carbohydrates, salts, nitrogen, trace elements, etc. The selection of growth medium and the starter culture that includes the inoculum of choice depends upon the product desired. Say, e.g. production of ethanol requires carbohydrate-rich molasses in the growth medium and invertase secreting yeast as inoculum. For production of cheese, lactic acid fermenting bacteria are added to the curdling milk in the fermenters. Examples of fermenting bacteria are *Lactobacillus, Enterococcus, Streptococcus,* etc., and are known for their beneficial effects on gut microbiota. These are typically used in inoculum as *starter culture* and fed into the fermenting food (e.g. milk for production of cheese). Characterization of the strain of inoculum (here in this case lactic acid-producing bacteria) is a norm as per the regulatory guidelines to ensure safety and maintain dietary consumption quality (Hill et al. 2017). The method of culturing microbes and obtaining commercially important products through the method of fermentation is briefly described by Fig. 11.

The major thrust areas at present are, namely the development of biodegradable polymers and clean green energy by creating biofuels using the tools available through white biotechnology. Since the use of plastics has greatly damaged the environment that does not require further elaboration, a shift towards the alternatives of synthetic non-biodegradable polymers has become imminent. Also, use of fossil fuels like petroleum is slowly being reduced due to its limited reserves. Various environmental protection agencies and government constituted panels have called for research looking into alternative fuel sources which are environment compatible.

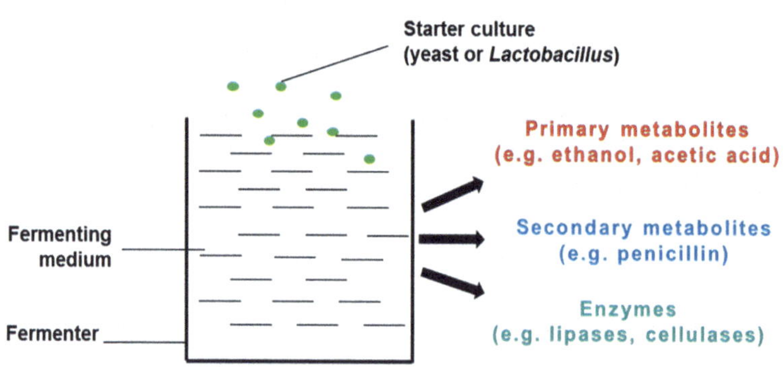

Fig. 11 Schematic describing the fermentation process and its products

Fig. 12 Switchgrass bunch grassland field (*Panicum virgatum*). Photo courtesy of Robert H. Mohlenbrock, hosted by the USDA-NRCS PLANTS Database/USDA SCS. 1991. Southern wetland flora: Field office guide to plant species. South National Technical Center, Fort Worth

Such fuels when derived from biomass meaning any plant or animal material are designated as *biofuels* as per US Environment Information Agency (EIA 2018). The most commonly used biofuels are bioethanol and biodiesel. Bioethanol is conventionally produced from carbohydrate-rich corn syrup or molasses. Recently, there has been a tilt observed towards the production of ethanol from cellulosic biomass. Switchgrass (*Panicum virgatum*; family: Poaceae) is a widely growing perennial bunch grass in most of the USA as shown in Fig. 12. The US Department of Agriculture (USDA) has selected the plant for production of bioethanol derived from the plant biomass because of its ease of growing and high yield. The ethanol thus produced is currently being mixed with gasoline at a final concentration of 10% (E10 fuel) and is used for automobiles (EIA 2018).

Biodiesel is a liquid fuel which is currently promoted as an alternative to diesel for its usage in public transport and commercial vehicles. It is manufactured by vegetable oils and animal fats by esterification of fatty acids with an alcohol (Omidvarborna et al. 2014). The National Biodiesel Board of America describes biodiesel as mono-alkyl esters of fatty acids and designate as B100 in its purest form. There are other parts mixed forms of biodiesel available, like B20 where it is 20% biodiesel of the total fuel. Plants utilized for the purpose of producing biodiesel include soybean, jatropha, rapeseed, etc. *Salicornia bigelovii* is a halophyte growing in coastal shorelines of America that has been identified as biomass feedstock for generating biofuel. It can survive well on salty water and is therefore suitable for large-scale cultivation by irrigating with salt-rich water, a common industrial effluent (Bullis 2010). A similar application of biomass lies in production of bioplastics which uses vegetable oils, straw, woodchips and food waste (McMillan 2010). Starch-based thermoplastics are used for making drug capsules by pharmaceutical companies. These plastics are actually a blend of several chemicals like polyhydroxy alkanoates and polylactic acid which are endowed with better water resistance and improved mechanical properties. Starch plastics are also used in the manufacturing of films used for packaging of food items (Halley and Avérous 2014). Another biodegradable polymer polylactic acid

Fig. 13 Chinese brake fern (*Pteris vittata*); a hyperaccumulator of arsenic. Photo courtesy of https://suzandtell.wordpress.com/2015/02/18/the-chinese-ladder-fern/

is being produced by fermenting metabolically engineered strain *E. coli* for supply to industries manufacturing it. Bacteria like Citrobacter, Klebsiella and Enterobacter have been known to produce 1,3-propanediol which is used for manufacturing adhesives, solvents and antifreeze agents (Lee and Jang 2006a, b).

Another area of growing importance is the field of phytoremediation that involves the usage of plant for environmental clean-up. The industrial pollutants are discharged into land, water and air which include heavy metals, organic compounds and high salt-containing water. These have become a growing menace for the society as they are severely damaging the environment and leading to rise in diseases and deaths. Heavy metals like cadmium, mercury and lead have been found to percolate down the soil and reach groundwater table thus polluting it and making it unfit for drinking or agricultural purpose. The plants that are able to extract these metals are used for the purpose of phytoremediation of the polluted soil. Plants capable of accumulating heavy metals more than 1000 mg/kg dry weight are called as hyperaccumulators (Reeves 2003). *Thlaspi caerulescens* (family: Brassicaceae) is being successfully used as hyperaccumulator for phytoremediation of cadmium and zinc. Arsenic remains one of the biggest threats to groundwater that is used for direct drinking purpose. A Chinese fern *Pteris vittata* of family—Pteridaceae (commonly known as Chinese brake fern) was demonstrated to be a hyperaccumulator of arsenic (Ma et al. 2001) (Fig. 13).

5 Summary

The field of biotechnology has evolved a lot since prehistoric times to what we see it today and continues to evolve further. Improvization of pre-existing knowledge and groundbreaking discoveries in the past centuries has led to significant overhauling of our understanding of biological processes. The causative agents of many previously incurable diseases have now an identity, a method of diagnosis and a possible treatment regime due to which life expectancy has increased considerably from those times. Biotechnology has touched upon lives of people globally through its various defined and undefined branches. Seminal discoveries governing the basic process of life, their applications in the areas of health care, agriculture and environment have revolutionized the practice of science and adaptation of technology in an unprecedented manner.

It is reasonable to conjecture that given the magnitude of applications in societal welfare, the field of biotechnology will be going to be a major area of research and investment. However, the endeavours pertaining to the investment are to be taken up with certain precautions because many such areas need accumulation of enough body of work before final commercialization can be realized. Specifically, areas like blue biotechnology which holds immense possibility of discovery of several crucial drugs, food supplements, industrially required materials are first needed to be explored through deep research. Only thereafter, the true feasibility of the product on a commercial scale can be adjudged. Areas like red biotechnology have many promising inventions and discoveries in its basket. Gene editing tool like CRISPR method is hailed across the globe as a major breakthrough in the sciences that can cure several genetic diseases which are incurable till date. Application of the same in the field of green biotechnology can lead to the generation of crops and livestock with improved or superior traits for ensuring food and nutrition security of the public. Increasing pollution of the ecosystem has prompted the scientists to adapt these technologies for better environmental clean-up. The exploitation of phytoremediation for absorbing heavy metals and degradation of oil spills in the oceans is an exciting area of exploration. Increasing the industrial output by utilization of biologics or living organisms helps in generation of several biodegradable materials which avoids conventional hazardous chemical synthetic processes.

What remains to be done is an exhaustive list of tasks that needs to be taken up seriously to ensure better practice of biotechnology throughout its various branches. The formation of ethical guidelines for application of genetic techniques for embryo manipulation needs to be carefully drafted. This is essential to prevent the abuse of technology for personal fame and reputation at the cost of suffering of humanity. Release of GM crops with proper evaluation of pros and cons must form an integral part of crop improvement programs in poor and developing countries. The area of blue biotechnology is still in its infancy, and therefore, more impetus is needed to attract people towards exploring marine genetic resources. Marine resource exploitation should be carried out with the sense of responsibility without damaging the ecosystem. The concept of sustained development needs to be implemented in essence so as to propel further research while ensuring judicious use of the natural resources.

References

Aizpurua-Olaizola O, Navarro P et al (2016) Microencapsulation and storage stability of polyphenols from Vitis vinifera grape wastes. Food Chem 190:614–621

Aldridge S (1999) The Discovery and Development of Penicillin 1928–1945: The Alexander Fleming Laboratory Museum, London, UK, November 19, 1999: an International Historic Chemical Landmark, American Chemical Society and Royal Society of Chemistry

Anderson NL, Anderson NG (1998) Proteome and proteomics: new technologies, new concepts, and new words. Electrophoresis 19(11):1853–1861

Andre G, Pellegrini M, Pellegrini L (2002) Cosmetic or dermatological compositions especially useful as antiaging and antiwrinkle products comprise extracts of Ascophyllum and Helopteris seaweeds. I. N. D. L. P. Industrielle, France

Aoi T, Yae K et al (2008) Generation of pluripotent stem cells from adult mouse liver and stomach cells. Science 321(5889):699–702

Aranovich D, Lewinsohn E, Zaccai M (2007) Post-harvest enhancement of aroma in transgenic lisianthus (*Eustoma grandiflorum*) using the *Clarkia breweri* benzyl alcohol acetyl transferase (BEAT) gene. Postharvest Biol Biotechnol 43(2):255–260

Arber W, Linn S (1969) DNA modification and restriction. Annu Rev Biochem 38:467–500

Arrieta JM, Arnaud-Haond S et al (2010) What lies underneath: conserving the oceans' genetic resources. Proc Natl Acad Sci U S A 107(43):18318–18324

Banerjee A, Sharma R et al (2002) *Botryococcus braunii*: a renewable source of hydrocarbons and other chemicals. Crit Rev Biotechnol 22(3):245–279

Banjara M, Zhu L, Shen G et al (2012) Expression of an *Arabidopsis* sodium/proton antiporter gene (AtNHX1) in peanut to improve salt tolerance. Plant Biotechnol Rep 6(1)

Barghouthi SA (2011) A universal method for the identification of bacteria based on general PCR primers. Indian J Microbiol 51(4):430–444

Barnett JA (2003) Beginnings of microbiology and biochemistry: the contribution of yeast research. Microbiology 149(Pt 3):557–567

Bartlett JM, Stirling D (2003) A short history of the polymerase chain reaction. Methods Mol Biol 226:3–6

Belakova J, Horynova M et al (2007) DNA vaccines: are they still just a powerful tool for the future? Arch Immunol Ther Exp (Warsz) 55(6):387–398

Benn CS, Netea MG et al (2013) A small jab—a big effect: nonspecific immunomodulation by vaccines. Trends Immunol 34(9):431–439

Blunt JW, Copp BR et al (2013) Marine natural products. Nat Prod Rep 30(2):237–323

Boyhan D, Daniell H (2011) Low-cost production of proinsulin in tobacco and lettuce chloroplasts for injectable or oral delivery of functional insulin and C-peptide. Plant Biotechnol J 9(5):585–598

Brown M, Wittwer C (2000) Flow cytometry: principles and clinical applications in hematology. Clin Chem 46(8 Pt 2):1221–1229

Bud R (1993) The uses of life: a history of biotechnology. Cambridge University Press, Cambridge; New York

Bull, A. T., Holt, G., and Lilly, M. D. (1982). Biotechnology: International Trends and Perspectives. Organisation for Economic Co-Operation and Development, Paris.

Bullis K (2010) Biofuels from saltwater crops. from https://www.technologyreview.com/s/417390/biofuels-from-saltwater-crops/. 5 Feb 2010

CDC (2017) National diabetes statistics report. C. f. D. C. a. Prevention, National Institute of Diabetes and Digestive and Kidney Diseases (NIH): 20

Chagastelles PC, Nardi NB (2011) Biology of stem cells: an overview. Kidney Int Suppl (2011) **1**(3): 63–67

Chiang JL, Kirkman MS et al (2014) Type 1 diabetes through the life span: a position statement of the American Diabetes Association. Diabetes Care 37(7):2034–2054

Choi T, Huang M et al (1991) A generic intron increases gene expression in transgenic mice. Mol Cell Biol 11(6):3070–3074

Clive J (2015) Executive summary: ISAAA Brief 46-2013. I. S. f. t. A. o. A.-B. Applications, ISAAA
Cole RD (1996) Choh Hao Li: April 21, 1913—November 28, 1987. Biogr Mem Natl Acad Sci 70:221–239
Collins MW, Bryant JA et al (2007) Design and information in biology: from molecules to systems. WIT Press, Southampton, UK; Boston
Consortium TWTCC (2007) Genome-wide association study of 14,000 cases of seven common diseases and 3,000 shared controls. Nature 447(7145):661–678
Convention on Biological Diversity (1992).
Crick FHC, Watson JD (1954) The complementary structure of deoxyribonucleic acid. Proc R Soc Lond A 223(1152)
Culver KW, Morgan RA et al (1990) In vivo expression and survival of gene-modified T lymphocytes in rhesus monkeys. Hum Gene Ther 1(4):399–410
Cyranoski D, Reardon S (2015) Embryo editing sparks epic debate. Nat News 520(7549):593–595
D'Amour KA, Bang AG et al (2006) Production of pancreatic hormone-expressing endocrine cells from human embryonic stem cells. Nat Biotechnol 24(11):1392–1401
Danna K, Nathans D (1971) Specific cleavage of simian virus 40 DNA by restriction endonuclease of Hemophilus influenzae. Proc Natl Acad Sci U S A 68(12):2913–2917
Das S, Pingle MR et al (2008) Detection and serotyping of dengue virus in serum samples by multiplex reverse transcriptase PCR-ligase detection reaction assay. J Clin Microbiol 46(10):3276–3284
DaSilva E (2004) The colours of biotechnology: science, development and humankind. Electron J Biotechnol 7(3)
Daston L, Lunbeck E (2011) Histories of scientific observation. University of Chicago Press, Chicago
Doughman SD, Krupanidhi S et al (2007) Omega-3 fatty acids for nutrition and medicine: considering microalgae oil as a vegetarian source of EPA and DHA. Curr Diabetes Rev 3(3):198–203
Eaves CJ (2015) Hematopoietic stem cells: concepts, definitions, and the new reality. Blood 125(17):2605–2613
ECORYS (2014) Study in support of impact assessment work on blue biotechnology. ECORYS, London/Berlin/Brussels/Rotterdam
Edward Jenner (1749–1823) History
Edwards J (1999) Why dolly matters: kinship, culture and cloning. Ethnos 64(3–4):301–324
EIA U (2018) Biofuels: ethanol and biodiesel explained. From https://www.eia.gov/energyexplained/index.cfm?page=biofuel_home. 8 Aug 2018
ERA-NET, EMB, MB (2017) Marine biotechnology: advancing innovation in Europe's bioeconomy. European Marine Board (EMB), Oostende, Belgium
Erickson B (2009) FDA approves drug from transgenic goat milk. Chemical and Engineering News. American Chemical Society, Washington, DC
Ernst HC (2014) On the extension of the germ theory to the etiology of certain common diseases. University of Adelaide, South Australia
Fiechter ABT (2000) History of modern biotechnology. Springer Science & Business Media, Berlin
Frazzetto G (2003) White biotechnology. EMBO Rep 4(9):835–837
Friedmann T, Roblin R (1972) Gene therapy for human genetic disease? Science 175(4025):949–955
Gasiunas G, Barrangou R et al (2012) Cas9-crRNA ribonucleoprotein complex mediates specific DNA cleavage for adaptive immunity in bacteria. Proc Natl Acad Sci U S A 109(39):E2579–E2586
Gaynes R (2017) The discovery of penicillin—new insights after more than 75 years of clinical use. Emerg Infect Dis 23(5):5
Genentech (1993) Genentech receives FDA approval to market Nutropin for growth failure due to chronic renal insufficiency. Press Release, from https://www.gene.com/media/press-releases/4393/1993-11-17/genentech-receives-fda-approval-to-marke

Gerstein MB, Bruce C et al (2007) What is a gene, post-ENCODE? History and updated definition. Genome Res 17(6):669–681
Gilbert DA, Reid YA et al (1990) Application of DNA fingerprints for cell-line individualization. Am J Hum Genet 47(3):499–514
Goncalves GAR, Paiva RMA (2017) Gene therapy: advances, challenges and perspectives. Einstein (Sao Paulo) 15(3):369–375
Good RA, Meuwissen HJ et al (1969) Bone marrow transplantation: correction of immune deficit in lymphopenic immunologic deficiency and correction of an immunologically induced pancytopenia. Trans Assoc Am Physicians 82:278–285
Gros F (1992) The gene civilization. McGraw Hill, New York
Guionet A, Hosseini B et al (2017) A new mechanism for efficient hydrocarbon electro-extraction from *Botryococcus braunii*. Biotechnol Biofuels 10(1):39
Haines JL, Hauser MA et al (2005) Complement factor H variant increases the risk of age-related macular degeneration. Science 308(5720):419–421
Halley PJ, Avérous L (2014) Starch polymers: from genetic engineering to green applications
Helden AV (2010) The origins of the telescope. Amsterdam University Press
Heux S, Meynial-Salles I et al (2015) White biotechnology: state of the art strategies for the development of biocatalysts for biorefining. Biotechnol Adv 33(8):1653–1670
Hicks DG, Tubbs RR (2005) Assessment of the HER2 status in breast cancer by fluorescence in situ hybridization: a technical review with interpretive guidelines. Hum Pathol 36(3):250–261
Hicks DJ, Fooks AR et al (2012) Developments in rabies vaccines. Clin Exp Immunol 169(3):199–204
Hill D, Sugrue I et al (2017) Recent advances in microbial fermentation for dairy and health. F1000Res 6:751
Hirschhorn R, Vawter GF et al (1979) Adenosine deaminase deficiency: frequency and comparative pathology in autosomally recessive severe combined immunodeficiency. Clin Immunol Immunopathol 14(1):107–120
Hsu PD, Lander ES et al (2014) Development and applications of CRISPR-Cas9 for genome engineering. Cell 157(6):1262–1278
Hu W, Kaminski R et al (2014) RNA-directed gene editing specifically eradicates latent and prevents new HIV-1 infection. Proc Natl Acad Sci U S A 111(31):11461–11466
Hughes SS (2011) Genentech: the beginnings of biotech. University of Chicago Press, Chicago; London
Ishino Y, Shinagawa H et al (1987) Nucleotide sequence of the iap gene, responsible for alkaline phosphatase isozyme conversion in *Escherichia coli*, and identification of the gene product. J Bacteriol 169(12):5429–5433
Ivancevic TD, Kurtovic NK et al (2014) The role of immunophenotyping in differential diagnosis of chronic lymphocytic leukemia. Srp Arh Celok Lek 142(3–4):197–203
Jakab K, Gopcsa L et al (2005) Application of flow cytometry immunophenotyping and multidrug resistance assay in B-cell acute lymphoid leukemia and multiple myeloma. Neoplasma 52(1):36–42
Jansen R, Embden JD et al (2002) Identification of genes that are associated with DNA repeats in prokaryotes. Mol Microbiol 43(6):1565–1575
Jinek M, Chylinski K et al (2012) A programmable dual-RNA-guided DNA endonuclease in adaptive bacterial immunity. Science 337(6096):816–821
Kafarski P (2012) Rainbow code of biotechnology. CHEMIK 66(8):811–816
Keating GM, Noble S (2003) Recombinant hepatitis B vaccine (Engerix-B): a review of its immunogenicity and protective efficacy against hepatitis B. Drugs 63(10):1021–1051
Kogler G, Sensken S et al (2004) A new human somatic stem cell from placental cord blood with intrinsic pluripotent differentiation potential. J Exp Med 200(2):123–135
Koreth J, O'Leary JJ et al (1996) Microsatellites and PCR genomic analysis. J Pathol 178(3):239–248

Kubista M, Andrade JM et al (2006) The real-time polymerase chain reaction. Mol Aspects Med 27(2–3):95–125
Kues WA, Niemann H (2011) Advances in farm animal transgenesis. Prev Vet Med 102(2):146–156
Kuiper K (2010) The Britannica guide to theories and ideas that changed the modern world. Britannica Educational Pub. in association with Rosen Educational Services, New York, NY
Kuleshova L, Gianaroli L et al (1999) Birth following vitrification of a small number of human oocytes: case report. Hum Reprod 14(12):3077–3079
Laflamme MA, Gold J et al (2005) Formation of human myocardium in the rat heart from human embryonic stem cells. Am J Pathol 167(3):663–671
Lakshmi V, Sudha T et al (2011) Application of polymerase chain reaction to detect HIV-1 DNA in pools of dried blood spots. Indian J Microbiol 51(2):147–152
Landgraf W, Sandow J (2016) Recombinant human insulins—clinical efficacy and safety in diabetes therapy. Eur Endocrinol 12(1):12–17
Landmarks, ACS, NHC (2009) Deciphering the genetic code. Commemorative Booklet
Lane N (2015) The unseen world: reflections on Leeuwenhoek (1677) 'Concerning little animals'. Philos Trans R Soc Lond B Biol Sci 370(1666)
Lansdown AB (2002) Calcium: a potential central regulator in wound healing in the skin. Wound Repair Regen 10(5):271–285
Lee SY, Jang SH (2006a) Commentaries & analyses—WHITE BIOTECHNOLOGY. Asia-Pacific Biotech News 10(10):559–563
Lee SY, Jang SH (2006b) WHITE BIOTECHNOLOGY—commentaries & analyses. Asia-Pacific Biotech News 10(10)
Liang P, Jin LH et al (2006) Human neural stem cells promote corticospinal axons regeneration and synapse reformation in injured spinal cord of rats. Chin Med J (Engl) 119(16):1331–1338
Liang P, Xu Y et al (2015) CRISPR/Cas9-mediated gene editing in human tripronuclear zygotes. Protein Cell 6(5):363–372
Liu Z, Cai Y et al (2018) Cloning of macaque monkeys by somatic cell nuclear transfer. Cell 172(4):881–887, e887
Lonnerdal B, Iyer S (1995) Lactoferrin: molecular structure and biological function. Annu Rev Nutr 15:93–110
Lorence A, Verpoorte R (2004) Gene transfer and expression in plants. Methods Mol Biol 267:329–350
Lorenz MG, Wackernagel W (1994) Bacterial gene transfer by natural genetic transformation in the environment. Microbiol Rev 58(3):563–602
Ma LQ, Komar KM et al (2001) A fern that hyperaccumulates arsenic. Nature 411(6836):438
Maksimenko OG, Deykin AV et al (2013) Use of transgenic animals in biotechnology: prospects and problems. Acta Naturae 5(1):33–46
Mancinelli L, Cronin M et al (2000) Pharmacogenomics: the promise of personalized medicine. AAPS PharmSci 2(1):E4
Mavilio F, Ferrari G (2008) Genetic modification of somatic stem cells. The progress, problems and prospects of a new therapeutic technology. EMBO Rep 9(Suppl 1):S64–S69
McIntosh M, Cruz LJ et al (1982) Isolation and structure of a peptide toxin from the marine snail *Conus magus*. Arch Biochem Biophys 218(1):329–334
McLaren A (2000) Cloning: pathways to a pluripotent future. Science 288(5472):1775–1780
McMillan JD (2010) Biotechnology for fuels and chemicals. In: The thirty-first symposium: proceedings of the thirty-first symposium on biotechnology for fuels and chemicals, 3–6 May 2009, San Francisco, CA. New York, Springer
Milstein C (1999) The hybridoma revolution: an offshoot of basic research. BioEssays 21(11):966–973
Nester EW, Gordon MP et al (1984) Crown Gall: a molecular and physiological analysis. Ann Rev Plant Physiol 35(1):387–413
Nykiforuk CL, Boothe JG et al (2006) Transgenic expression and recovery of biologically active recombinant human insulin from *Arabidopsis thaliana* seeds. Plant Biotechnol J 4(1):77–85

Oliveira SC, Rosinha GM et al (1999) Immunological properties of gene vaccines delivered by different routes. Braz J Med Biol Res 32(2):207–214

Omidvarborna H, Kumar A et al (2014) Characterization of particulate matter emitted from transit buses fueled with B20 in idle modes. J Environ Chem Eng 2(4):2335–2342

Parvez S, Malik KA et al (2006) Probiotics and their fermented food products are beneficial for health. J Appl Microbiol 100(6):1171–1185

Pasteur L (1879) Studies on fermentation; the diseases of beer, their causes, and the means of preventing them. Macmillan and Co., London

Pearson S, Jia H et al (2004) China approves first gene therapy. Nat Biotechnol 22(1):3–4

Perucho M (1996) Cancer of the microsatellite mutator phenotype. Biol Chem 377(11):675–684

Piechulla B, Effmert U (2010) Biosynthesis and regulation of flower scent. Plant developmental biology—biotechnological perspectives. In: Davey M, Pua E (eds) Springer Nature Switzerland AG, Berlin, Heidelberg

Plotkin S (2014) History of vaccination. Proc Natl Acad Sci U S A 111(34):12283–12287

Pope JE, Deer TR (2013) Ziconotide: a clinical update and pharmacologic review. Expert Opin Pharmacother 14(7):957–966

Poste G (2001) Molecular diagnostics: a powerful new component of the healthcare value chain. Expert Rev Mol Diagn 1(1):1–5

Quispe-Huamanquispe DG, Gheysen G et al (2017) Horizontal gene transfer contributes to plant evolution: the case of agrobacterium T-DNAs. Front Plant Sci 8:2015

Rd D (2004) Recombinant human antithrombin III: rhATIII. Drugs R D 5(2):110–112

Reeves RD (2003) Tropical hyperaccumulators of metals and their potential for phytoextraction. Plant Soil 249(1):57–65

Ribeiro BD, Coelho MAZ et al (2016) Principles of green chemistry and white biotechnology. In: White biotechnology for sustainable chemistry, pp 1–8

Riedel S (2005) Edward Jenner and the history of smallpox and vaccination. Proc (Bayl Univ Med Cent) 18(1):21–25

Rinehart KL (2000) Antitumor compounds from tunicates. Med Res Rev 20(1):1–27

Rocha-Munive MG, Soberon M et al (2018) Evaluation of the impact of genetically modified cotton after 20 years of cultivation in Mexico. Front Bioeng Biotechnol 6:82

Roewer L (2013) DNA fingerprinting in forensics: past, present, future. Investig Genet 4(1):22

Rollier CS, Reyes-Sandoval A et al (2011) Viral vectors as vaccine platforms: deployment in sight. Curr Opin Immunol 23(3):377–382

Rosenberg SA, Aebersold P et al (1990) Gene transfer into humans–immunotherapy of patients with advanced melanoma, using tumor-infiltrating lymphocytes modified by retroviral gene transduction. N Engl J Med 323(9):570–578

Ruane J, Sonnino A et al (2006) The role of biotechnology in exploring and protecting agricultural genetic resources. Food and Agriculture Organization of the United Nations, Rome

Russell N, Bessell E et al (2000) Allogeneic haemopoietic stem cell transplantation for multiple myeloma or plasma cell leukaemia using fractionated total body radiation and high-dose melphalan conditioning. Acta Oncol 39(7):837–841

Saad R (2005) Discovery, development, and current applications of DNA identity testing. Proc (Bayl Univ Med Cent) 18(2):130–133

Sabirova JS, Ferrer M et al (2006) Mutation in a "tesB-like" hydroxyacyl-coenzyme A-specific thioesterase gene causes hyperproduction of extracellular polyhydroxyalkanoates by *Alcanivorax borkumensis* SK2. J Bacteriol 188(24):8452–8459

Sadelain M (2006) Recent advances in globin gene transfer for the treatment of beta-thalassemia and sickle cell anemia. Curr Opin Hematol 13(3):142–148

Sambu S (2015) A Bayesian approach to optimizing cryopreservation protocols. PeerJ 3:e1039

Schneiker S, Martins dos Santos VA et al (2006) Genome sequence of the ubiquitous hydrocarbon-degrading marine bacterium *Alcanivorax borkumensis*. Nat Biotechnol 24(8):997–1004

Schumann K, Lin S et al (2015) Generation of knock-in primary human T cells using Cas9 ribonucleoproteins. Proc Natl Acad Sci U S A 112(33):10437–10442

Setlow JK, Hollaender A (2002) Genetic engineering; principles and methods, vol 4, Plenum Press, New York
Sharma AK, Bury MI et al (2011) A nonhuman primate model for urinary bladder regeneration using autologous sources of bone marrow-derived mesenchymal stem cells. Stem Cells 29(2):241–250
Shata MT, Stevceva L et al (2000) Recent advances with recombinant bacterial vaccine vectors. Mol Med Today 6(2):66–71
Singh RS (2018) Introduction to principles of plant pathology. Oxford and IBH Publishing
Slater A, Scott NW et al (2008) Plant biotechnology: the genetic manipulation of plants. Oxford University Press, Oxford; New York
Smith CA, O'Maille G et al (2005) METLIN: a metabolite mass spectral database. Ther Drug Monit 27(6):747–751
Spemann H (1938) Embryonic development and induction. Yale University Press; H. Milford, Oxford University Press, New Haven, London
Spencer CC, Su Z et al (2009) Designing genome-wide association studies: sample size, power, imputation, and the choice of genotyping chip. PLoS Genet 5(5):e1000477
Sussulini A (2017) Metabolomics: from fundamentals to clinical applications
Tabebordbar M, Zhu K et al (2016) In vivo gene editing in dystrophic mouse muscle and muscle stem cells. Science 351(6271):407–411
Takahashi N, Miura I et al (2005) Fluorescence in situ hybridization monitoring of BCR-ABL-positive neutrophils in chronic-phase chronic myeloid leukemia patients during the primary stage of imatinib mesylate therapy. Int J Hematol 81(3):235–241
Tang G, Qin J et al (2009) Golden rice is an effective source of vitamin A. Am J Clin Nutr 89(6):1776–1783
Tautz D (1993) Notes on the definition and nomenclature of tandemly repetitive DNA sequences. EXS 67:21–28
Tempest HG, Ko E et al (2008) Sperm aneuploidy frequencies analysed before and after chemotherapy in testicular cancer and Hodgkin's lymphoma patients. Hum Reprod 23(2):251–258
Thomson JA, Itskovitz-Eldor J et al (1998) Embryonic stem cell lines derived from human blastocysts. Science 282(5391):1145–1147
TMR (2018, Jan 2018) Human growth hormone market: ToC. Market research report, from https://www.transparencymarketresearch.com/report-toc/16283
US F&DA (1982) Human insulin receives FDA approval. FDA Drug Bull 12(3):18–19
Valones MA, Guimaraes RL et al (2009) Principles and applications of polymerase chain reaction in medical diagnostic fields: a review. Braz J Microbiol 40(1):1–11
van Berkel PH, Welling MM et al (2002) Large scale production of recombinant human lactoferrin in the milk of transgenic cows. Nat Biotechnol 20(5):484–487
Vogenberg FR, Isaacson Barash C et al (2010) Personalized medicine: part 1: evolution and development into theranostics. P T 35(10):560–576
Vutyavanich T, Piromlertamorn W et al (2010) Rapid freezing versus slow programmable freezing of human spermatozoa. Fertil Steril 93(6):1921–1928
Walker L, Levine H et al (2006) Koch's postulates and infectious proteins. Acta Neuropathol 112(1):1–4
Walsh G (2005) Therapeutic insulins and their large-scale manufacture. Appl Microbiol Biotechnol 67(2):151–159
Waltz E (2017) First genetically engineered salmon sold in Canada. Scientific American, Springer Nature America Inc., United States of America
Whitelaw CB, Archibald AL et al (1991) Targeting expression to the mammary gland: intronic sequences can enhance the efficiency of gene expression in transgenic mice. Transgenic Res 1(1):3–13
WHO (2018) Factsheet: diabetes. W. H. Organization
Wishart DS, Tzur D et al (2007) HMDB: the human metabolome database. Nucleic Acids Res 35(Database issue):D521–D526

Wu CH, Huang H et al (2002) The protein information resource: an integrated public resource of functional annotation of proteins. Nucleic Acids Res 30(1):35–37

Yamamoto Y (2002) PCR in diagnosis of infection: detection of bacteria in cerebrospinal fluids. Clin Diagn Lab Immunol 9(3):508–514

Yang NS, Burkholder J et al (1990) In vivo and in vitro gene transfer to mammalian somatic cells by particle bombardment. Proc Natl Acad Sci U S A 87(24):9568–9572

Yang P, Wang J et al (2008) Cattle mammary bioreactor generated by a novel procedure of transgenic cloning for large-scale production of functional human lactoferrin. PLoS ONE 3(10):e3453

Yaskowiak ES, Shears MA et al (2006) Characterization and multi-generational stability of the growth hormone transgene (EO-1alpha) responsible for enhanced growth rates in Atlantic Salmon. Transgenic Res 15(4):465–480

Zhao JR, Bai YJ et al (2005) Detection of hepatitis B virus DNA by real-time PCR using TaqMan-MGB probe technology. World J Gastroenterol 11(4):508–510

Zhou S, Flamier A et al (2015) Differentiation of human embryonic stem cells into cone photoreceptors through simultaneous inhibition of BMP, TGFbeta and Wnt signaling. Development 142(19):3294–3306

Harsh Kumar did his doctorate from Regional Centre for Biotechnology, Faridabad (India) in 2019 and is a cell biology enthusiast. His doctoral work demonstrated novel biochemical and functional crosstalk between early endocytosis and exocytosis-related proteins for facilitating the abscission process in cytokinesis. He is currently working as a postdoctoral research associate at the Genetics Division, Department of Pediatrics, All India Institute of Medical Sciences, New Delhi (India).

Tiny Targets, BIG Impact!

CRISPR: The Revolutionary Gene Editing Tool with Far-Reaching Applications

Sohinee Bhattacharyya and Anindit Mukherjee

Abstract In the realm of scientific research, there is an immense value for biological research tools, which can modify, insert, and delete DNA sequences of cells or organisms in order to understand the function of specific genes. One of the most revolutionary scientific revelations of the twenty-first century has been the development of the programmable CRISPR-Cas9 genome engineering technology. The groundbreaking CRISPR-Cas9 gene modulating system is derived from the original type II CRISPR-Cas system, whose primary purpose was to endow bacteria with adaptive immunity to viral infections. CRISPR-associated protein 9 or simply Cas9 is an enzyme that belongs to the class endonuclease by function. In brief, Cas9 functions to introduce double-stranded breaks in the target DNA sequence by using a guide RNA sequence to form base pairs with target DNA sequences. The elegance and simplicity of the CRISPR-Cas9 genome editing system combined with its cost-effectiveness and efficiency in precisely targeting and editing genomic sequences have heralded a transformative phase in basic biological research. Scientists around the globe are using this technology for novel groundbreaking applications that range from improving human therapeutics, improving disease resistance of industrially important crops, and for basic biology research. This chapter aims to present the key milestones in the evolution of this revolutionary CRISPR-Cas9 genome editing tool and will also discuss the far-reaching impacts of this transformative technology in the domains of basic science, biotechnology, and medicine.

Keywords CRISPR · Cas9 · TALEN · Homologous recombination · Genome engineering · Gene editing

---

S. Bhattacharyya (✉)
Oregon Health and Science University, Portland, USA
e-mail: bhattacs@ohsu.edu

A. Mukherjee
Oregon State University, Corvallis, USA

A. Saxena (ed.), *Biotechnology Business - Concept to Delivery*, EcoProduction,
https://doi.org/10.1007/978-3-030-36130-3_2

1 Development of the CRISPR-Cas9 Genome Editing Tool

The human genome sets the stage for the blueprint of life. The accurate and precise transfer of genetic information from gene to messenger RNA and ultimately to protein (the central dogma of molecular biology) is indispensible in preventing the development of genetic anomalies and diseases (Crick 1970). Rare events where fidelity of genetic information transfer is compromised may lead to the development of serious genetic diseases. Some starring examples include genetically inherited BRCA1 mutations in breast cancer patients and in chronic myeloid leukemia patients, somatic BCR-ABL1 fusions. These diseases are a couple of examples where mutations in crucial genes have been directly linked to the development of human pathologies. As a matter of fact, although strategies to target human diseases often are directed toward inhibiting metabolic pathways or cellular signaling pathways, the most efficacious therapeutic regimens often target the underlying genetic anomalies. Scientists have been developing technologies to edit underlying genomic abnormalities (such as mutations), and technologies for manipulating genetic information (DNA) hold tremendous potential for treating human genetic diseases, even certain cancers. In order to correct underlying genetic abnormalities, scientists are now using harnessing the power of genome engineering tools to introduce specifically targeted modifications in the gene pool of living cells. The path to the discovery of genetic tools that could introduce site-specific modifications to the genetic code was not an easy path. It took scientists around the world several years of hard work and research to discover the mechanism behind the revolutionary genetic engineering breakthrough, the CRISPR/Cas9 system. Early developments in the field of genome engineering lead to the discovery of site-directed "zinc finger nucleases (ZFNs), TAL effector nucleases (TALENs)," which relied on the principles of "DNA–protein recognition" (Carroll 2011). However, due to the laborious efforts involved in protein engineering, both these systems had severe drawbacks in large-scale studies (Carroll 2011). Finally, after years of research, a breakthrough in the scientific field was the discovery of the "CRISPR/Cas9 system." Compared to the older gene editing technologies, the CRISPR/Cas9 system offered the researchers site specificity, versatility, high efficiency, and ease of use. By simply programming guide RNA sequences (guide RNAs direct the Cas9 endonuclease), Cas9 endonuclease can be specifically directed to edit defined DNA sequences.

The process of genome engineering involves the introduction of targeted modifications to the genome. There is a growing demand for technologies that can achieve this easily and efficiently in mammalian cells as technologies that can easily modify, delete, and insert DNA sequences hold immense potential to transform the realms of biotechnology, basic science, and therapy. Genomes of eukaryotic organisms are complex in nature and contain billions of DNA bases. This complex structural organization makes it immensely difficult to introduce new modifications into the existing DNA sequence. One of the earliest methods used for genome targeting has been the use of homologous recombination (Capecchi 1989). "Homologous recombination" is a biological process that is widely used by cells to repair double-stranded breaks, and it can also be used to integrate external repair sequences that are sequentially homologous to the donor site. Homologous recombination-mediated genome editing downstream of targeted DNA double-stranded breaks was demonstrated very elegantly in a group of pioneering studies by Rudin et al. (1989). Following this, Carroll and Chandrasegaran developed zinc finger protein designer nucleases for precise, sequence-specific homologous recombination (Bibikova et al. 2001, 2002, 2003). However, one of the major drawbacks of this process is that homologous recombination occurs extremely rarely (1 in 106–109 cells) (Capecchi 1989). This presents an enormous challenge to the adaptation of this method for large-scale applications. To overcome this obstacle, scientists developed four principal classes of elegant genome editing tools that could introduce locus-specific double-stranded breaks in the DNA: (I)Meganucleases—originally coming from mobile genetic elements of microbes, introns, or inteins (Smith et al. 2006). Meganucleases are functionally classified as homing endonucleases which may be further broadly classified as intein endonucleases and intron endonucleases. Mobile genetic elements such a sintrons propagate by intervening at a specific locus in the genomic sequence, and the expression of the meganuclease introduces a cleavage in the corresponding intron free allele. This initial cleavage event leads to the duplication of the intron sequence at the cleavage site by double-stranded DNA break repair (homologous recombination). (II) Zinc finger (ZF) nucleases—this class of nucleases were developed by artificially fusing a DNA cleavage functional domain to a "zinc finger DNA-binding domain." This class of nucleases can be engineered to modify specific genetic sequences and rely on eukaryotic transcription factors (Urnov et al. 2005; Miller et al. 2007). (III) "Transcription activator-like effectors (TALEs)"—TAL effectors are secreted proteins that are derived from Xanthomonas bacteria (Miller et al. 2011; Christian et al. 2010; Boch et al. 2009; Cong et al. 2013) when they infect plants. TALENs or "Transcription Activator-Like Effector Nucleases" are a group of restriction enzymes that can be modified to target specific genetic sequences. They are artificially synthesized by structurally fusing a TAL effector DNA-binding domain to a functional DNA cleavage domain. (IV) The "RNA-guided DNA endonuclease Cas9—derived from bacterial adaptive immune system CRISPR" (Cong et al. 2013; Mali et al. 2013).

Meganuclease, ZF, and TALE proteins exploit the same principle; they bind to specific DNA sequences through specific protein–DNA interactions. However, they have severe drawbacks that make it challenging for use in genome engineering. For example, meganucleases, ZF nucleases, and TALEs lack precise specificity to their target DNA sequences and strategies to overcome this limitation can render the method expensive and labor-intensive. Given these challenges, scientists worked toward the development of a simpler and more precise nuclease. The CRISPR story started gaining momentum 1987 (Ishino et al. 1987) when a set of 29 nucleotide repeats downstream of the *iap* gene (the *iap* gene encodes for an enzyme which is involved in isozyme conversion of alkaline phosphatase in *E. coli*) was reported. As more microbial genomes were deciphered, it was discovered that unique sets of clustered repeat elements are present in greater than 40% of all bacteria sequenced and a vast majority of archaea (Mojica et al. 2005). Despite this compelling finding, the precise function of these repeat elements was unknown until 2005, when through investigations of the nonrepetitive spacer sequences separating the repeat elements suggested that they were derived from bacteriophages (Pourcel et al. 2005; Bolotin et al. 2005) These findings coupled to earlier studies that elucidated that bacteriophages are unable to infect bacterial cells harboring spacer sequences that correspond to their own genomic sequence, led to the possibility that "CRISPR arrays" function as microbial immunity modules and act to mediate viral defense. The first concrete proof that the CRISPR array is a mechanism of antiviral defense came from the studies with the bacterial strain *Streptococcus thermophilus*, which is routinely used in dairy cultures; they demonstrated a nucleic-acid-based functional module in which "CRISPR spacers" regulate preciseness of sequence selectivity, while Cas enzymes regulate acquisition of spacer sequences and develop immunity to the phage (Barrangou et al. 2007).

Around this time, two revolutionary studies elucidated the key components that are required for efficient genome editing by the "CRISPR system." In the first study, Moineau and colleagues (Barrangou et al. 2007) elucidated Cas9 as the only nuclease that mediates sequence-specific DNA cleavage (Garneau et al. 2010). In the second study, Deltcheva and colleagues revealed the existence of a crucial component of the CRISPR module—a "trans-activating crRNA (tracrRNA)" that forms a hybrid complex with crRNA to facilitate RNA-guided trafficking of Cas9 to selected genomic sequences (Deltcheva et al. 2011).

2 Domain Organization of the "CRISPR/Cas9 System"

The "CRISPR-Cas9 technology" originally evolved from "type II CRISPR-Cas systems," whose primary function was to bestow upon bacteria immunity from infecting viruses (Carroll 2011; Cong et al. 2013; Mali et al. 2013; Jinek et al. 2012; Cho et al. 2013; Cox et al. 2015; Wright et al. 2016; Doudna and Charpentier 2014; Wiedenheft et al. 2012; Fineran and Charpentier 2012; Horvath and Barrangou 2010). In bacteria, CRISPR-Cas9 loci are composed of an array of identical CRISPR repeats and an operon (a set of linked genes with the capacity to regulate other genes) that encodes the Cas nuclease complex (Carroll 2011; Cong et al. 2013; Mali et al. 2013; Jinek et al. 2012; Cho et al. 2013; Cox et al. 2015; Wright et al. 2016; Doudna and Charpentier 2014; Wiedenheft et al. 2012; Fineran and Charpentier 2012; Horvath and Barrangou 2010; Ran et al. 2013). The CRISPR nuclease Cas9 achieves target specificity with the help of a short guide RNA that hybridizes to and cuts target DNA via Watson–Crick base pairing. Typically, the CRISPR RNA guide sequences match phage genomic sequences in order to render immunity to the bacteriophages. However, one huge advantage of the CRISPR system is that these targeting sequences can be easily switched with a selected genomic sequence of interest to reassign the Cas9 nuclease. Upon invasion by viruses and plasmids, adaptive immunity in conferred in three key steps: (i) A short sequence of the invading DNA is acquired and inserted into the CRISPR loci, (ii) maturation of precursor crRNA to generate functional RNA duplex, tracrRNA:crRNA duplexes with invader targeting sequences, and (iii) site-specific double-stranded break in the DNA by Cas proteins at sites corresponding to the RNA duplex, tracrRNA:crRNA guide sequence (Carroll 2011; Cong et al. 2013; Mali et al. 2013; Jinek et al. 2012; Wright et al. 2016; Doudna and Charpentier 2014; Wiedenheft et al. 2012; Fineran and Charpentier 2012; Horvath and Barrangou 2010; Ran et al. 2013). Very recently, scientists engineered the dual tracrRNA:crRNA to a single-guide RNA. This novel and elegant engineering advance generated an easy-to-use bipartite system (Doudna and Charpentier 2014). The easily programmable Cas9 targeting system, its high accuracy, and precise sequence specificity as a sequence-specific nuclease have opened up a diverse range of biological applications that range from basic research to biotechnology and medicine.

3 Revolutionary Applications of the CRISPR/Cas9 System: Genome Engineering and Beyond

3.1 Generation of Knockout Cell Lines for Basic Science Research

The easy-to-use "CRISPR-Cas9 genome editing system" has made the generation of genetic knockout cells very feasible and efficient for a diverse array of cell systems including cancer-derived patient organoids, primary immune cells as well as stem cells. Generating such knockouts is invaluable to the advancement of scientific and clinical research as it allows researchers to rapidly establish functional roles of their genes of interest (Jinek et al. 2013; Grobarczyk et al. 2015; Matano et al. 2015; Hou et al. 2013; Drost et al. 2015; Mandal et al. 2014) (Fig. 1).

4 Rapid Generation of Animal Models

Aside from being invaluable in the field of cell biology and basic science research, CRISPR-Cas9 system has drastically enhanced the efficiency of generating animal models of human pathologies (Shen et al. 2013; Li et al. 2013a, c; Wang et al. 2013). CRISPR-Cas9 system can now be used to generate mice with genetic alterations in the desired gene of interest in order to ultimately understand the functions of the genes involved. This can be achieved by using electroporation or microinjection of zygotes instead of using the traditional expensive and labor-intensive procedure of embryonic stem cell (ESC) manipulation (Shen et al. 2013; Li et al. 2013a, c; Wang et al. 2013).

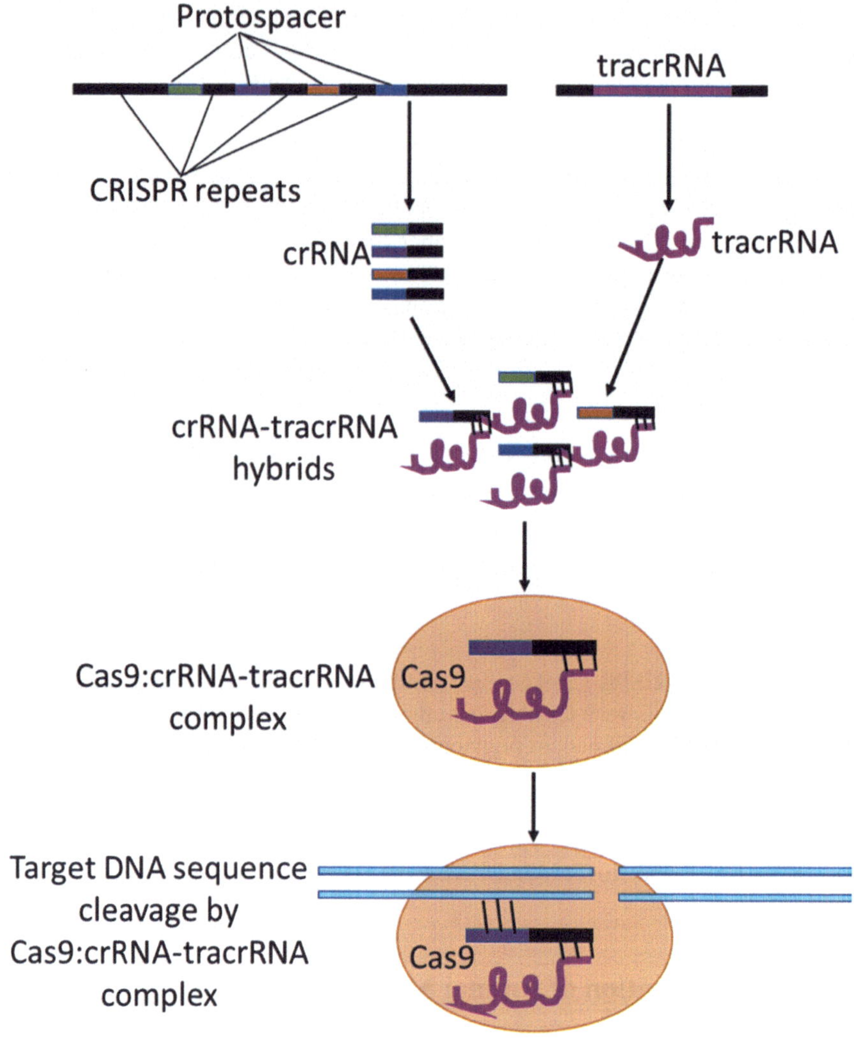

Fig. 1 Mechanism of action of CRIPSR Cas9 module (i) a short sequence of the invading DNA is incorporated into the CRISPR loci, (ii) maturation of precursor crRNA to generate functional RNA duplex, tracrRNA:crRNA duplexes with invader targeting sequences, and (iii) site-specific double-stranded break in the DNA by Cas proteins at sites complementary to the RNA duplex, tracrRNA:crRNA guide sequence

5 Using CRISPR-Cas to Make Novel Therapeutic Approaches

Apart from its tremendous potential in enhancing the knowledge base of the genetic basis of human malignancies, the power of the "CRISPR-Cas9 system" can be harnessed to develop novel therapeutic approaches. A starring example is the use of CRISPR-Cas to engineer CAR T cells that have tremendous promise in the area of cancer immunotherapy (Schumann et al. 2015). These genetically engineered T cells, which express specific tumor-targeting receptors, have demonstrated tremendous therapeutic potential for lymphoma and leukemia patients (Schumann et al. 2015).

Another excellent example to demonstrate the clinical efficacy of the "CRISPR-Cas9 system" is the treatment of patients with haemoglobinopathies like sickle cell disease (SCD). Sickle cell disease is caused by a mutation in the Hemoglobin gene. Recently, SCD patients have been subjected to ex vivo gene correction using CRISPR-Cas9 genome modification (Mandal et al. 2014; Hoban et al. 2015; DeWitt et al. 2016). This mutational event causes a shift in the amino acids coded. More specifically, substitution of a glutamate (Glu) to valine (Val) is the causal genetic mutation that is found in all sickle cell disease patients. CRISPR-Cas9 has been able to correct this mutation by using a single-stranded oligonucleotide introduction or by using an integrase defective lentiviral vector (Mandal et al. 2014; Hoban et al. 2015; DeWitt et al. 2016).

6 "CRISPR/Cas9 System" in Genetically Modifying Plants and Cattle Strains

Recently, scientists around the world have started using the "CRISPR-Cas system" to genetically modify agricultural crops and cattle strains in order to produce more economically beneficial, improved strains. For example, in economically important agricultural crops like rice, wheat, sweet orange, and sorghum, target genes such as genes that make the plants susceptible to microbial pathogenic infections have been edited out using "CRISPR-Cas9 technology" (Li et al. 2013b; Nekrasov et al. 2013; Xie and Yang 2013; Jiang et al. 2013; Upadhyay et al. 2013). As an added advantage, it was recently reported that these sites directed changes introduced by "CRISPR-Cas9" did not have any detrimental off target effects and importantly; these newly introduced genetic edits were inherited by the next generation of plants without the development of new spontaneous and random mutational events. Thus, "CRISPR-Cas9" genome editing tool is also set to revolutionize the agriculture and dairy industry by creating new disease resistance strains that are genetically protected from microbial diseases and pests (Li et al. 2013b; Nekrasov et al. 2013; Xie and Yang 2013; Jiang et al. 2013; Upadhyay et al. 2013).

7 Conclusions

To summarize, the "CRISPR/Cas9 system" is poised to revolutionize the fields of drug discovery and targeted therapy by enabling the precise and specific modulation of genetic information in disease models. In the near future, the power of the CRISPR/Cas9 system will enable scientists and clinicians to introduce corrective mutations or modify gene regulatory elements or correct gene-splicing patterns.

References

Barrangou R, Fremaux C, Deveau H, Richards M, Boyaval P, Moineau S, Romero DA, Horvath P (2007) Science 315:1709–1712
Bibikova M, Beumer K, Trautman JK, Carroll D (2003) Science 300:764
Bibikova M, Carroll D, Segal DJ, Trautman JK, Smith J, Kim YG, Chandrasegaran S (2001) Mol Cell Biol 21:289–297
Bibikova M, Golic M, Golic KG, Carroll D (2002) Genetics 161:1169–1175
Boch J, Scholze H, Schornack S, Landgraf A, Hahn S, Kay S, Lahaye T, Nickstadt A, Bonas U (2009) Science 326:1509–1512
Bolotin A, Quinquis B, Sorokin A, Ehrlich SD (2005) Microbiology 151:2551–2561
Capecchi MR (1989) Science 244:1288–1292
Carroll D (2011) Genetics 188:773–782
Cho SW, Kim S, Kim JM, Kim JS (2013) Nat Biotechnol 31:230–232
Christian M, Cermak T, Doyle EL, Schmidt C, Zhang F, Hummel A, Bogdanove AJ, Voytas DF (2010) Genetics 186:757–761
Cong L, Ran FA, Cox D, Lin S, Barretto R, Habib N, Hsu PD, Wu X, Jiang W, Marraffini LA, Zhang F (2013) Science 339:819–823
Cox DB, Platt RJ, Zhang F (2015) Nat Med 21:121–131
Crick F (1970) Nature 227:561–563
Deltcheva E, Chylinski K, Sharma CM, Gonzales K, Chao Y, Pirzada ZA, Eckert MR, Vogel J, Charpentier E (2011) Nature 471:602–607
DeWitt MA, Magis W, Bray NL, Wang T, Berman JR, Urbinati F, Heo SJ, Mitros T, Munoz DP, Boffelli D, Kohn DB, Walters MC, Carroll D, Martin DI, Corn JE (2016) Sci Transl Med 8:360ra134
Doudna JA, Charpentier E (2014) Science 346:1258096
Drost J, van Jaarsveld RH, Ponsioen B, Zimberlin C, van Boxtel R, Buijs A, Sachs N, Overmeer RM, Offerhaus GJ, Begthel H, Korving J, van de Wetering M, Schwank G, Logtenberg M, Cuppen E, Snippert HJ, Medema JP, Kops GJ, Clevers H (2015) Nature 521:43–47
Fineran PC, Charpentier E (2012) Virology 434:202–209
Garneau JE, Dupuis ME, Villion M, Romero DA, Barrangou R, Boyaval P, Fremaux C, Horvath P, Magadan AH, Moineau S (2010) Nature 468:67–71
Grobarczyk B, Franco B, Hanon K, Malgrange B (2015) Stem Cell Rev 11:774–787
Hoban MD, Cost GJ, Mendel MC, Romero Z, Kaufman ML, Joglekar AV, Ho M, Lumaquin D, Gray D, Lill GR, Cooper AR, Urbinati F, Senadheera S, Zhu A, Liu PQ, Paschon DE, Zhang L, Rebar EJ, Wilber A, Wang X, Gregory PD, Holmes MC, Reik A, Hollis RP, Kohn DB (2015) Blood 125:2597–2604
Horvath P, Barrangou R (2010) Science 327:167–170
Hou Z, Zhang Y, Propson NE, Howden SE, Chu LF, Sontheimer EJ, Thomson JA (2013) Proc Natl Acad Sci U S A 110:15644–15649
Ishino Y, Shinagawa H, Makino K, Amemura M, Nakata A (1987) J Bacteriol 169:5429–5433

Jiang W, Zhou H, Bi H, Fromm M, Yang B, Weeks DP (2013) Nucleic Acids Res 41:e188
Jinek M, Chylinski K, Fonfara I, Hauer M, Doudna JA, Charpentier E (2012) Science 337:816–821
Jinek M, East A, Cheng A, Lin S, Ma E, Doudna J (2013) Elife 2:e00471
Li D, Qiu Z, Shao Y, Chen Y, Guan Y, Liu M, Li Y, Gao N, Wang L, Lu X, Zhao Y, Liu M (2013a) Nat Biotechnol 31:681–683
Li JF, Norville JE, Aach J, McCormack M, Zhang D, Bush J, Church GM, Sheen J (2013b) Nat Biotechnol 31:688–691
Li W, Teng F, Li T, Zhou Q (2013c) Nat Biotechnol 31:684–686
Mali P, Yang L, Esvelt KM, Aach J, Guell M, DiCarlo JE, Norville JE, Church GM (2013) Science 339:823–826
Mandal PK, Ferreira LM, Collins R, Meissner TB, Boutwell CL, Friesen M, Vrbanac V, Garrison BS, Stortchevoi A, Bryder D, Musunuru K, Brand H, Tager AM, Allen TM, Talkowski ME, Rossi DJ, Cowan CA (2014) Cell Stem Cell 15:643–652
Matano M, Date S, Shimokawa M, Takano A, Fujii M, Ohta Y, Watanabe T, Kanai T, Sato T (2015) Nat Med 21:256–262
Miller JC, Holmes MC, Wang J, Guschin DY, Lee YL, Rupniewski I, Beausejour CM, Waite AJ, Wang NS, Kim KA, Gregory PD, Pabo CO, Rebar EJ (2007) Nat Biotechnol 25:778–785
Miller JC, Tan S, Qiao G, Barlow KA, Wang J, Xia DF, Meng X, Paschon DE, Leung E, Hinkley SJ, Dulay GP, Hua KL, Ankoudinova I, Cost GJ, Urnov FD, Zhang HS, Holmes MC, Zhang L, Gregory PD, Rebar EJ (2011) Nat Biotechnol 29:143–148
Mojica FJ, Diez-Villasenor C, Garcia-Martinez J, Soria E (2005) J Mol Evol 60:174–182
Nekrasov V, Staskawicz B, Weigel D, Jones JD, Kamoun S (2013) Nat Biotechnol 31:691–693
Pourcel C, Salvignol G, Vergnaud G (2005) Microbiology 151:653–663
Ran FA, Hsu PD, Wright J, Agarwala V, Scott DA, Zhang F (2013) Nat Protoc 8:2281–2308
Rudin N, Sugarman E, Haber JE (1989) Genetics 122:519–534
Schumann K, Lin S, Boyer E, Simeonov DR, Subramaniam M, Gate RE, Haliburton GE, Ye CJ, Bluestone JA, Doudna JA, Marson A (2015) Proc Natl Acad Sci U S A 112:10437–10442
Shen B, Zhang J, Wu H, Wang J, Ma K, Li Z, Zhang X, Zhang P, Huang X (2013) Cell Res 23:720–723
Smith J, Grizot S, Arnould S, Duclert A, Epinat JC, Chames P, Prieto J, Redondo P, Blanco FJ, Bravo J, Montoya G, Paques F, Duchateau P (2006) Nucleic Acids Res 34:e149
Upadhyay SK, Kumar J, Alok A, Tuli R (2013) G3 (Bethesda) 3:2233–2238
Urnov FD, Miller JC, Lee YL, Beausejour CM, Rock JM, Augustus S, Jamieson AC, Porteus MH, Gregory PD, Holmes MC (2005) Nature 435:646–651
Wang H, Yang H, Shivalila CS, Dawlaty MM, Cheng AW, Zhang F, Jaenisch R (2013) Cell 153:910–918
Wiedenheft B, Sternberg SH, Doudna JA (2012) Nature 482:331–338
Wright AV, Nunez JK, Doudna JA (2016) Cell 164:29–44
Xie K, Yang Y (2013) Mol Plant 6:1975–1983

Sohinee Bhattacharyya In the Sherman Lab at OHSU, Sohinee is presently working on the gene regulatory interaction networks that exist between pancreatic tumor and stroma. She joined the University of Nebraska Medical Center in 2009 to pursue her Ph.D. in pathology and microbiology where she investigated the novel role of the endocytic recycling protein EHD1 in mouse early embryonic morphogenesis. After completing her Ph.D., she joined the University of Pittsburgh School of Medicine as a postdoctoral fellow. Here, under the guidance of Dr. Ora Weisz, she investigated the cellular signaling pathways that regulate apical endocytosis in proximal tubule cells in response to fluid shear stress.

Anindit Mukherjee received his Ph.D. degree in cellular and integrative physiology from the University of Nebraska Medical Center in 2012. He is presently working at Oregon State University as an expert in epithelial biology and is currently focused on treatment of cystic fibrosis with the help of nanoparticle-delivered therapeutics.

Nanotechnology in Medicine

Anindit Mukherjee and Sohinee Bhattacharyya

Abstract Nanotechnology is at the frontier of medicine. It plays a critical role in expanding diversity and reach of the drug or drug-like molecules in patients. Nanoparticles act as a protected transport vehicle for hydrophobic drugs or therapeutics such as mRNA, siRNA, and DNA which are normally not bioavailable. In this chapter, we review the different kinds of non-viral nanocarriers that are being employed in modern-day medicine. We also discuss the challenges to effective delivery and function of these particles as well as their application in various aspects of disease treatment.

Keywords Nanotechnology · DNA · RNA · Liposome · Polyplex · Dendrimer · Quantum dot · Metal nanoparticle · Endocytosis · Macropinocytosis · Phagocytosis · Clathrin · Caveolin · Therapeutics · Niemann–Pick type C · Fabry · AIDS · HIV · Cancer

1 Introduction

Nanoparticles are engineered nanoscale structures that are increasingly being used in the field of biomedical science for targeting and delivery of therapeutics. Most of these particles are less than 100 nm in size. Nanomaterials have been used as a protective vehicle to increase solubility and stability of therapeutics in the bloodstream. This ability has helped expand the repertoire of cargo such as hydrophobic drugs, DNA, RNA, peptides, and antibodies that are normally labile in the blood stream, to be used as therapeutics. Additionally, particles have been engineered to take advantage of biological and cellular processes for enhancing targeting and delivery of cargo into the intended cells or tissues. However, the additional challenge for nanoparticle engineering is to design the particles that are at once capable of protecting their cargo in the blood stream but once inside the cells they will fall apart to release their cargo.

A. Mukherjee (✉)
Oregon State University, Corvallis, USA
e-mail: mukherja@ohsu.edu

S. Bhattacharyya
Oregon Health and Science University, Portland, USA

A. Saxena (ed.), *Biotechnology Business - Concept to Delivery*, EcoProduction,
https://doi.org/10.1007/978-3-030-36130-3_3

2 Types of Nanoparticles and Their Application in the Biomedical Field

A wide variety of materials have been engineered to be used as nanoparticles. They include polymers, lipids, inorganic particles, viral vectors, cell ghosts, etc. In this review, we will focus on the synthetic particles, namely liposomes, nanoparticle albumin-bound (nab) technology, polymeric nanoparticles, dendrimers, metal nanoparticles, and molecular-targeted nanoparticles.

Liposomes or lipid nanoparticles or lipoplexes are single-layered or multilayered spherical lipid vesicles with an aqueous core. Liposomes are made with natural or synthetic lipids and are capable of packaging both hydrophilic and hydrophobic cargoes. Some of the earliest of attempts to introduce exogenous genes into cells were performed using lipid-based particles. By the 1980s, various genetic materials such as DNA and RNA were introduced into cells by utilizing the fusogenic potential of liposomes. Once inside the endosome, the liposomes destabilize and fuse with the endosomal membrane in an inside-out manner and release their cargo into the cell cytoplasm. Several liposomal formulations are approved for clinical applications. Cationic lipids such as *N*-[1-(2,3-dioleyloxy)propyl]-*N*,*N*,*N*-trimethylammonium chloride (DOTMA) have been frequently used for formulating lipoplexes. Variations of these formulations such as ones using ionizable lipids such as Dlin-MC3-DMA (MC3) or novel ionizable lipid (lipid 5) have been successful in greatly increasing intracellular nucleic acid delivery in cell and animal models. Liposomes can be formulated with polymers such as polyethylene glycol (PEG) or with fusogenic lipids such as dioleoylphosphatidyl-ethanolamine (DOPE) and cholesterol for increasing transfection efficacy. Cholesterol helps in cholesterol-dependent mechanisms for cellular entry of the nanoparticles, whereas adding polymers such as PEG to the surface increases stability of the particles in serum. Nanoformulation of doxorubicin, an anticancer drug, has resulted in markedly increased bioavailability (Torchilin 2005).

Protein nanoparticles made with albumin-bound technology (nab) have made use of capacity of albumin to carry and release hydrophobic cargo. Additionally, albumin is transcytosed via glycoprotein 60 receptor which helps in intracellular delivery of the cargo. Paclitaxel, a hydrophobic drug, is made soluble using which nab technology is sold by the name Abraxane (Benson et al. 1978).

Polymeric nanoparticles are made with biocompatible block co-polymers consisting of two or more block polymer chains with differing hydrophilicity. The hydrophobic block forms the core, while the hydrophilic block forms the outer shell that interacts with the aqueous phase. Various therapeutics can be packaged in the core of these particles for delivery (Zhang et al. 2008).

Dendrimers are well-defined branched monodispersed polymers that can trap molecules within their core. Dendrimers can also be conjugated with therapeutic or diagnostic molecules. Most dendrimer formulation is 1–20 nm in size. These particles are made with polymers such as polyamidoamine (PAMAM), melamine, poly l-glutamic acid (PG), polyethyleneimine, or natural elements such as amino acids, sugars, and nucleotides. Polyamidoamine (PAMAM), melamine, poly l-glutamic acid (PG), polyethyleneimine consist of a core, branched units emerging from the core in layers, and an outer layer with repeat units of functional end groups. The solvent-filled core forms the nanoscale cargo hold, while the defined exterior surface allows it to interact with biological domains (Svenson and Tomalia 2005).

Metal nanoparticle are synthesized using biocompatible inert metals such as gold, nickel, and silica. Different shapes such as nanoshell or quantum dots or nanorods have been fashioned with metals. Unique optical and magnetic properties of these nanoparticles have been variously used for diagnostic and imaging purposes, e.g., MRI contrast imaging. Supermagnetic nanoparticles made with maghemite, cobalt ferrite, etc., that oscillate between a strongly magnetic and a non-magnetic state in the presence and absence of a magnetic field, respectively, are being studied for in vivo applications. Oscillating magnetic field causes these particles to generate heat, which is being used for intracellular hyperthermia-based applications.

Unique optical properties of semiconductor-based quantum dots are also being extensively being used for biological applications, particularly in the field of imaging. Several properties of these materials including their ability to absorb and emit quantized packets of energy, the wide range of absorption spectrum, a large difference between the absorption and emission spectra and harnessing the discrete and resistance to photobleaching have made them very attractive for biological imaging application.

Metal nanoparticles have also been conjugated with hydrophobic drug for better dispersal and delivery. For example, gold nanoparticles have been coated with amphiphilic thiols, whose hydrophobic pockets' package take up the hydrophobic therapeutics such as tamoxifen, and the hydrophilic group makes the particles water-soluble (Minelli et al. 2010).

Molecular-targeted nanoparticles have their surface decorated with targeting molecules such as monoclonal antibodies, aptamers, peptides, and small molecules (such as folate) against cell-specific markers. These molecules bind to specific marker proteins on cells and get internalized via receptor-mediated endocytosis. Targeted nanoparticles are attracting a lot of attention. Formulations such as BIND-014, a polymeric nanoparticle formulation of docetaxel, with controlled release property went into clinical trial (Wang et al. 2008).

3 Barriers to Nanoparticle Delivery

Cellular uptake mechanisms at once present challenges and opportunities for nanoparticle-based intracellular delivery mechanisms. Cells take up materials from the extracellular milieu using a variety of mechanisms broadly known as endocytosis (Sahay et al. 2010). Endocytosis can be subdivided into phagocytosis, a process by which cells engulf large particles up to 20 μm in diameter and pinocytosis, a more generalized form of endocytosis, involving uptake of solutes along with extracellular fluids (Conner and Schmid 2003). Phagocytosis is primarily observed in specialized phagocytic cells such as macrophages, neutrophils, monocytes, and dendritic cells, whereas pinocytosis is observed in all cells. Depending on the molecular mechanism of uptake, pinocytosis can be primarily of four types: clathrin-mediated, macropinocytosis, caveolin-mediated, and clathrin-caveolin-independent. Of which, the last three together are known as clathrin-independent mechanisms. Clathrin-independent uptake mechanisms involve receptors and membrane-associated proteins other than clathrin, such as caveolin, ARf6, Flotillin, CDC42, and RhoA that aid in forming outside-in membrane vesicles for uptake of extracellular materials.

Calthrin-mediated endocytosis is a hallmark of all cells, while clathrin-independent mechanisms display some cell-type bias. For example, polarized epithelial cells lack caveolae on their apical membrane (Lahtinen et al. 2003; Vogel et al. 1998; De et al. 2014), whereas caveolae formation is prevalent in muscle, endothelium, and fibroblasts (Doherty and McMahon 2009).

Depending on the cargo, cell type, and the vesicle type, majority of the endocytosed materials are either degraded via lysosomal pathway, transcytosed out of the cell or recycled back to the cells surface. Only a small percentage gets delivered into the cytosol via a mysterious mechanism, broadly termed as endosomal escape.

Nanoparticle-based intracellular delivery strategies have tried to take advantage of the heterogeneity of uptake mechanism by changing composition, size, shape, and charge of the particles to target them to specific endocytic compartments in order to enhance endosomal escape (Varkouhi et al. 2011). Additionally, clathrin-independent mechanisms have been targeted to enhance particle delivery to non-epithelial cells (Oh et al. 2007).

4 Application of Nanoparticles in Diseases

Lysosomal storage disorder: Lysosomal storage disorders (LSDs) are an attractive nanoparticle-based therapeutic target, while small molecule screen in HeLa cells identified several effectors such as V-ATPase, arachidonic acid metabolism, and Cathepsin that affect delivery of LNP to lysosomes (Sahay et al. 2013). The inherent defects in intracellular trafficking result in accumulation of material inside the late endosomes and lysosomes. This type of accumulation has been utilized for enhancing the effectiveness of nanoparticle delivery.

Niemann–Pick disease type C: Niemann–Pick Type C (NPC) is an inherited autosomal recessive lysosomal storage disorder that affects approximately 1:120,000 children globally and results in hepatomegaly, progressive neurodegeneration, and finally death (Futerman and van Meer 2004; Vanier 2010). The disease is caused by mutation in NPC1 and 2 that result in cholesterol accumulation in the endosomes and blocks efflux out of the late endosomal compartments (Ikonen 2008). Superparamagnetic iron oxide nanoparticles (SPIONs)-based ferrofluid nanoparticles were used to isolate late endosomes and study the difference in the composition of lysosomal glycocalyx (Tharkeshwar et al. 2017; Kosicek et al. 2018). Increase in retention time and defective degradation machinery of endosomes facilitated endosomal escape of siRNA into the cytosol (Sahay et al. 2013; Eltoukhy et al. 2014). Interestingly, PEG-based lipid nanoparticles have shown to be a bioactive agent by facilitating cholesterol solubilization and efflux from these compartments (Brown et al. 2016).

In Niemann–Pick type B, caused due to acid sphingomyelinase deficiency, clathrin-mediated endocytosis is impaired (Rappaport et al. 2014). ICAM1-conjugated nanocarriers were used to deliver acid sphingomyelinase to the lysosome by clathrin-independent route by utilizing ICAM1-dependent route and decreased sphingomyelin storage in mouse lung (Garnacho et al. 2008; Muro et al. 2003, 2005).

In Fabry disease caused due to α-galactosidase A (α-Gal) deficiency and in Pompe disease caused due to acid α-glucosidase (GAA) deficiency, nanoparticle was used to α-Gal and GAA, to reduce lysosomal accumulation of globotriaosylceramide (Gb3) and to decrease excessive glycogen storage, respectively, in mice (Hsu et al. 2011, 2012).

AIDS: Nanoparticles have been used to tackle particularly challenging targets such as HIV virus. HIV infects and hides inside patient's immune cells such as peripheral monocytes and macrophages. The virus thus uses the immune system as a protective reservoir as well as a vehicle for transporting and spreading the virus to different parts of the body (Amiji et al. 2009; Aquaro et al. 2002). Although saquinavir is very effective in reducing virus load, it is highly hydrophobic which makes for a key hurdle in using it for effective therapy. Poly(ethylene oxide)-modified poly(epsilon-caprolactone) (PEO-PCL) and polyhexylcyanoacrylate-encapsulated saquinavir nanoparticles were used to overcome the challenge of poor solubility and increase intracellular delivery of the drug and effective reduction of HIV infection (Bender et al. 1996; Shah and Amiji 2006).

Cancer: The use of nanotechnology for treatment of cancer is an active area of biomedical research. Nanoparticles are particularly useful in solubilizing several cancer drugs that are mostly hydrophobic molecules which short plasma half-life. Moreover, solid tumors are characterized by leaky vasculature that arises due to rapid and dysregulated proliferation of endothelial cells as well as due to decreased abundance of pericytes. This unique leakiness allows nanocarriers along with their cargo to be differentially enriched in the tumor microenvironment and not be cleared through the lymphatic system as would be the case in normal tissue. This enrichment is termed as enhanced permeability retention (EPR) effect. Polyethylene glycol (PEG)-coated doxorubicin, a hydrophobic drug, has been coated to enhance solubility, increase circulation time, enhance EPR, and decrease non-specific toxicity of the drug (Barenholz 2012; Unezaki et al. 1995). Doxorubicin nanoparticles have been widely used to treat different types of cancer, including Kaposis sarcoma, ovarian cancer lung carcinoma, prostate cancer, etc. (Sakakibara et al. 1996; Northfelt et al. 1996; Stürzl et al. 1994; Vaage et al. 1997; Muggia et al. 1997; Huang et al. 1992). Albumin-bound nanoparticle (Nab) technology has been used to encapsulate the hydrophobic antitumor agent, paclitaxel to enhance albumin driven endocytosis via the cell surface receptor GP60 and SPARC (secreted protein, acidic, and rich in cysteine), a protein that is upregulated in multiple tumors (Tai and Tang 2008; Tiruppathi et al. 1997; Minshall et al. 2000). This product is commercially available as Abraxane. Docetaxel, another anticancer drug, was encapsulated in a controlled release poly(lactide-co-glycolide)-poly(ethylene glycol) (PLGA-PEG) polymer that is decorated with peptide ligands against prostate-specific membrane antigen (PSMA) that is overexpressed in certain prostate cancers. The product named BIND-014 is currently in phase 2 trial (Autio et al. 2018; Hoff et al. 2016). Drugs conjugated with protein also form another important class of nanomedicine. Doxorubicin and Camptothecin conjugated with albumin and a cathepsin cleavable linker. Enzyme-catalyzed cleavage of the linker in the lysosomal compartment leads to release of the drug (Akinc and Battaglia 2013; Schmid et al. 2007). Similarly, antitumor drugs are conjugated to antibodies via low-pH hydrolysable or Cathepsin-like enzyme cleavable linker. The antibodies bind specific proteins on the tumor cell surface and are internalized along with the drug. The low endosomal pH or specific lysosomal enzymes then cleave the linker and release the drug into the cancer cells (Diamantis and Banerji 2016; Ducry and Stump 2010). Several nanodrugs, such as Ontak, Mylotarg, Adcetris, Kadcyla, and Vintafolide, designed on the basis of aforementioned principle are either approved or in various phases of clinical trials (Nabhan and Tallman 2002; Younes et al. 2010; Verma et al. 2012; Lorusso et al. 2012). Ontak, for example, is an interleukin-2 (IL-2)–diphtheria toxin fusion protein that is used to treat cutaneous T-cell lymphoma (CTCL). It is targeted against the CD25 subunit of the IL-2 receptor CTCL cell surface (Olsen et al. 2001).

Nanoparticle-based gene therapy has also been studied in treating various cancer types (Babu et al. 2016). Nanocarriers have been deployed to deliver tumor suppressor genes such as p53 or siRNA against oncogenes such as c-Myc/MDM2/VEGF, surviving, hypoxia-inducible factor 1α, and KSP (Ramesh et al. 2001, 2004). siRNA-based nanoparticles have also being studied (Vaishnaw et al. 2010; Young et al. 2016; Feng et al. 2017; Gillespie et al. 2015; Chen et al. 2010).

5 Conclusion

Nanotechnology has opened exciting frontiers in the treatment of intractable medical problems. While current research is full of tremendous promise, challenges remain in taking basic research from bench to bedside. However, newer engineering and advances in materials and molecular techniques hold tremendous potential for nanomedicine to truly take hold and deliver personalized medicine with lower side effects.

References

Akinc A, Battaglia G (2013) Cold Spring Harb Perspect Biol 5
Amiji MM, Vyas TK, Shah LK (2009) Discov Med 6:157–162
Aquaro S, Caliò R, Balzarini J, Bellocchi MC, Garaci E, Perno CF (2002) Antiviral Res 55:209–225

Autio KA, Dreicer R, Anderson J, Garcia JA, Alva A, Hart LL, Milowsky MI, Posadas EM, Ryan CJ, Graf RP, Dittamore R, Schreiber NA, Summa JM, Youssoufian H, Morris MJ, Scher HI (2018) JAMA Oncol 4:1344–1351
Babu A, Muralidharan R, Amreddy N, Mehta M, Munshi A, Ramesh R (2016) IEEE Trans Nanobiosci 15:849–863
Barenholz YC (2012) J Controlled Release 160:117–134
Bender AR, Briesen HV, Kreuter J, Duncan IB, Rübsamen-Waigmann H (1996) Antimicrob Agents Chemother 40:1467–1471
Benson ML, Thomson RG, Valli VE (1978) Can J Comp Med 42:368–369
Brown A, Patel S, Ward C, Lorenz A, Ortiz M, DuRoss A, Wieghardt F, Esch A, Otten EG, Heiser LM, Korolchuk VI, Sun C, Sarkar S, Sahay G (2016) Sci Rep 6:31750
Chen Y, Zhu X, Zhang X, Liu B, Huang L (2010) Mol Ther 18:1650–1656
Conner SD, Schmid SL (2003) Nature 422:37–44
De S, Kuwahara S, Saito A (2014) Membranes (Basel) 4:333–355
Diamantis N, Banerji U (2016) Br J Cancer 114:362–367
Doherty GJ, McMahon HT (2009) Annu Rev Biochem 78:857–902
Ducry L, Stump B (2010) Bioconjugate Chem 21:5–13
Eltoukhy AA, Sahay G, Cunningham JM, Anderson DG (2014) ACS Nano 8:7905–7913
Feng C, Wang T, Zhang Y, Qu K, Tang S (2017) Am J Med Sci 354:506–512
Futerman AH, van Meer G (2004) Nat Rev Mol Cell Biol 5:554–565
Garnacho C, Dhami R, Simone E, Dziubla T, Leferovich J, Schuchman EH, Muzykantov V, Muro S (2008) J Pharmacol Exp Ther 325:400–408
Gillespie DL, Aguirre MT, Ravichandran S, Leishman LL, Berrondo C, Gamboa JT, Wang L, King R, Wang X, Tan M, Malamas A, Lu Z-R, Jensen RL (2015) J Neurosurg 122:331–341
Hoff DDV, Mita MM, Ramanathan RK, Weiss GJ, Mita AC, LoRusso PM, Burris HA, Hart LL, Low SC, Parsons DM, Zale SE, Summa JM, Youssoufian H, Sachdev JC (2016) Clin Cancer Res 22:3157–3163
Hsu J, Serrano D, Bhowmick T, Kumar K, Shen Y, Kuo YC, Garnacho C, Muro S (2011) J Control Release 149:323–331
Hsu J, Northrup L, Bhowmick T, Muro S (2012) Nanomedicine 8:731–739
Huang SK, Mayhew E, Gilani S, Lasic DD, Martin FJ, Papahadjopoulos D (1992) Cancer Res 52:6774–6781
Ikonen E (2008) Nat Rev Mol Cell Biol 9:125–138
Kosicek M, Gudelj I, Horvatic A, Jovic T, Vuckovic F, Lauc G, Hecimovic S (2018) Mol Cell Proteomics 17:631–642
Lahtinen U, Honsho M, Parton RG, Simons K, Verkade P (2003) FEBS Lett 538:85–88
Lorusso PM, Edelman MJ, Bever SL, Forman KM, Pilat M, Quinn MF, Li J, Heath EI, Malburg LM, Klein PJ, Leamon CP, Messmann RA, Sausville EA (2012) J Clin Oncol Official J Am Soc Clin Oncol 30:4011–4016
Minelli C, Lowe SB, Stevens MM (2010) Small 6:2336–2357
Minshall RD, Tiruppathi C, Vogel SM, Niles WD, Gilchrist A, Hamm HE, Malik AB (2000) J Cell Biol 150:1057–1070
Muggia FM, Hainsworth JD, Jeffers S, Miller P, Groshen S, Tan M, Roman L, Uziely B, Muderspach L, Garcia A, Burnett A, Greco FA, Morrow CP, Paradiso LJ, Liang LJ (1997) J Clin Oncol 15:987–993
Muro S, Wiewrodt R, Thomas A, Koniaris L, Albelda SM, Muzykantov VR, Koval M (2003) J Cell Sci 116:1599–1609
Muro S, Gajewski C, Koval M, Muzykantov VR (2005) Blood 105:650–658
Nabhan C, Tallman MS (2002) Clin Lymphoma 2(Suppl 1)S19–S23
Northfelt DW, Martin FJ, Working P, Volberding PA, Russell J, Newman M, Amantea MA, Kaplan LD (1996) J Clin Pharmacol 36:55–63
Oh P, Borgstrom P, Witkiewicz H, Li Y, Borgstrom BJ, Chrastina A, Iwata K, Zinn KR, Baldwin R, Testa JE, Schnitzer JE (2007) Nat Biotechnol 25:327–337

Olsen E, Duvic M, Frankel A, Kim Y, Martin A, Vonderheid E, Jegasothy B, Wood G, Gordon M, Heald P, Oseroff A, Pinter-Brown L, Bowen G, Kuzel T, Fivenson D, Foss F, Glode M, Molina A, Knobler E, Stewart S, Cooper K, Stevens S, Craig F, Reuben J, Bacha P, Nichols J (2001) J Clin Oncol Official J Am Soc Clin Oncol 19:376–388
Ramesh R, Saeki T, Smyth Templeton N, Ji L, Stephens LC, Ito I, Wilson DR, Wu Z, Branch CD, Minna JD, Roth JA (2001) Mol Therapy 3:337–350
Ramesh R, Ito I, Saito Y, Wu Z, Mhashikar AM, Wilson DR, Branch CD, Roth JA, Chada S (2004) DNA Cell Biol 23:850–857
Rappaport J, Garnacho C, Muro S (2014) Mol Pharm 11:2887–2895
Sahay G, Alakhova DY, Kabanov AV (2010) J Control Release 145:182–195
Sahay G, Querbes W, Alabi C, Eltoukhy A, Sarkar S, Zurenko C, Karagiannis E, Love K, Chen D, Zoncu R, Buganim Y, Schroeder A, Langer R, Anderson DG (2013) Nat Biotechnol 31:653–658
Sakakibara T, Chen F-A, Kida H, Kunieda K, Cuenca RE, Martin FJ, Bankert RB (1996) Cancer Res 56:3743–3746
Schmid B, Chung D-E, Warnecke A, Fichtner I, Kratz F (2007) Bioconjug Chem 18:702–716
Shah LK, Amiji MM (2006) Pharm Res 23:2638–2645
Stürzl M, Zietz C, Eisenburg B, Goebel F-D, Gillitzer R, Hofschneider PH, Bogner JR (1994) Res Virol 145:261–269
Svenson S, Tomalia DA (2005) Adv Drug Deliv Rev 57:2106–2129
Tai IT, Tang MJ (2008) Drug Resist Updates 11:231–246
Tharkeshwar AK, Trekker J, Vermeire W, Pauwels J, Sannerud R, Priestman DA, Te Vruchte D, Vints K, Baatsen P, Decuypere JP, Lu H, Martin S, Vangheluwe P, Swinnen JV, Lagae L, Impens F, Platt FM, Gevaert K, Annaert W (2017) Sci Rep 7:41408
Tiruppathi C, Song W, Bergenfeldt M, Sass P, Malik AB (1997) J Biol Chem 272:25968–25975
Torchilin VP (2005) Nat Rev Drug Discov 4:145–160
Unezaki S, Maruyama K, Ishida O, Suginaka A, Hosoda J-I, Iwatsuru M (1995) Int J Pharm 126:41–48
Vaage J, Donovan D, Uster P, Working P (1997) Br J Cancer 75:482–486
Vaishnaw AK, Gollob J, Gamba-Vitalo C, Hutabarat R, Sah D, Meyers R, de Fougerolles T, Maraganore J (2010) Silence 1:14
Vanier MT (2010) Orphanet J Rare Dis 5:16
Varkouhi AK, Scholte M, Storm G, Haisma HJ (2011) J Control Release 151:220–228
Verma S, Miles D, Gianni L, Krop IE, Welslau M, Baselga J, Pegram M, Oh D-Y, Diéras V, Guardino E, Fang L, Lu MW, Olsen S, Blackwell K, E. S. Group (2012) New Eng J Med 367:1783–1791
Vogel U, Sandvig K, van Deurs B (1998) J Cell Sci 111(Pt 6):825–832
Wang AZ, Gu F, Zhang L, Chan JM, Radovic-Moreno A, Shaikh MR, Farokhzad OC (2008) Expert Opin Biol Ther 8:1063–1070
Younes A, Bartlett NL, Leonard JP, Kennedy DA, Lynch CM, Sievers EL, Forero-Torres A (2010) New Eng J Med 363:1812–1821
Young SWS, Stenzel M, Jia-Lin Y (2016) Crit Rev Oncol/Hematol 98:159–169
Zhang L, Gu FX, Chan JM, Wang AZ, Langer RS, Farokhzad OC (2008) Clin Pharmacol Ther 83:761–769

Food for All; Nutraceuticals

The Rise of Nutraceuticals: Overview and Future

Nitika Kapoor, Vijay Lakshmi Jamwal, Manish R. Shukla and Sumit G. Gandhi

Abstract Recent changes in lifestyle and dietary patterns, along with increased industrialization, rising pollution and work pressure, have led to increased susceptibility to a plethora of disorders such as hypertension, obesity, diabetes, inflammatory and autoimmune conditions, and cancers. Recent advancement in the medical sciences has considerably helped in treatment of these conditions; however, it has been recently realized that nutrition management may play a huge role in prevention as well as in treatment to some extent. The market is presently flooded with hundreds of nutraceuticals/food supplements that not only help to prevent and tackle such disease conditions but may also provide general health promotion. Nutraceuticals are food substances intended to be supplemented to regular diet and may contribute to general well-being, provide protection from diseases, or may ameliorate disease conditions. Nutraceuticals may contain vitamins, lipids, proteins, carbohydrates, minerals, herbs or herbal extracts, or other necessary nutrients, depending on their emphases and health claims. This chapter provides an overview of different categories of nutraceuticals, their health applications, and business outlook.

Keywords Amino acids and proteins · Fatty acids · Herbs · Minerals · Microalgae · Nutraceuticals · Probiotics · Prebiotics · Vitamins

N. Kapoor · V. L. Jamwal · S. G. Gandhi (✉)
Plant Biotechnology Division, CSIR-Indian Institute of Integrative Medicine, Canal Road, Jammu 180001, India
e-mail: sumit@iiim.ac.in

M. R. Shukla
Research and Development Center, Reliance Industries Limited, CU-127, TC-30-C, 2nd Floor, RCP Thane-Belapur Road, Ghansoli, Navi Mumbai 400701, India

A. Saxena (ed.), *Biotechnology Business - Concept to Delivery*, EcoProduction,
https://doi.org/10.1007/978-3-030-36130-3_4

1 Introduction

1.1 Defining Nutraceuticals

The term Nutraceutical was coined in 1989 by Stephen De Felice. He joined the terms nutrition and pharmaceuticals to describe how nutrition can be used for disease prevention (DeFelice 1995). Food or food additives that supplement the diet and provide health benefits are popularly known as nutraceuticals, functional foods, dietary supplements, etc. (Dureja et al. 2003). Nutraceutical products may have substances that are traditionally not consumed as food, but may have a positive physiological benefit such as prevention or treatment of disease condition (Singh 2013; Kalra 2003). Nutraceuticals thus lie somewhere between food and drugs, although they are generally not as strictly regulated as the conventional modern drugs (Gulati and Ottaway 2006; Pandey et al. 2010). As per various regulatory agencies, there are minor differences between the definitions of nutraceuticals, functional foods, and dietary supplements. For instance, functional foods are defined as foods cooked using scientific intelligence and providing the body with the requisite quantity of vitamins, proteins, carbohydrates, minerals, fats, etc., for its healthy survival (El Sohaimy 2012). However, if such a food helps in prevention or treatment of a disease, other than anemia, then it may be called as a nutraceutical. Dietary supplement as defined by The Dietary Supplement Health and Education Act of 1994 (DSHEA) of the USA includes products other than tobacco that may contain mineral, vitamin, amino acids, herbs, etc., in the form of a syrup, capsule, or pills (Zeisel 1999). Such supplements are not projected to be used as sole food item or as conventional food and may help to increase longevity, provide health benefits, or fulfill the special age-related dietary requirements, etc. (Gupta et al. 2018). However, since we consider these as minor regulatory variations, and this chapter intends to give the reader an overview of nutraceuticals, we have clubbed them together in a single term nutraceutical and may be used interchangeably in the text with other terms functional foods and dietary supplements. As such in some countries, these terms are treated synonymously.

1.2 Need for Nutraceuticals

Sedentary lifestyle patterns and unhealthy eating habits, along with increased industrialization, rising pollution, and work pressure, have led to increased susceptibility to a plethora of disorders/diseases such as hypertension, obesity, diabetes, joint and skeleton problems, inflammatory and autoimmune conditions, psycho-neurological issues, and cancers (Prasad et al. 2012) (Fig. 1). Additionally, the average lifespan has steadily increased over the past few decades, with a large subset of adult population who suffer from such lifestyle-related disorders/diseases. Vegetarian diet and food rich in carbohydrates and fats may lead to protein deficiency (Elorinne et al. 2016). Protein deficiency is associated with fatty liver disease, loss of muscle mass, skin and

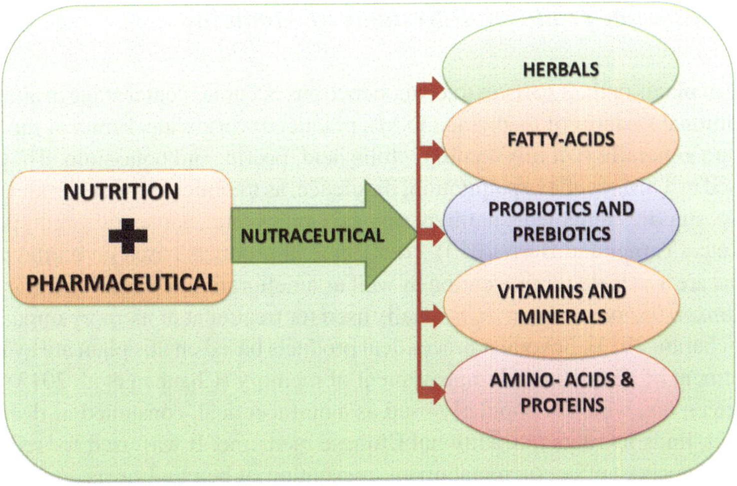

Fig. 1 Nutraceuticals and its major classes

hair problems, greater risk of bone fractures, and increased severity of infections. Increase in consumption of processed and fast food may also lead to high cholesterol and high blood sugar, which may eventually lead to cardiovascular diseases, obesity, and diabetes (Pereira et al. 2004; Swinburn et al. 2004; Bahadoran et al. 2015). Further, such diets may also be deficient in fiber, leading to difficult bowel movement and constipation (Cencic and Chingwaru 2010). Fiber-deficient diet has also been linked to diverticular disease (Ünlü et al. 2012), heart disease (McRae 2017), diabetes (Schultz et al. 2004), and inflammation (Krishnamurthy et al. 2012). Low-grade chronic inflammation, on the other hand, has been associated with heart disease, cancer, rheumatoid arthritis, diabetes, obesity, etc. (Coussens and Werb 2002). Obesity and diabetes are on the rise even in younger population and children (Pozzilli et al. 2011). Further, recent trends show an increase in vitamin and mineral deficiency, not only in economically weaker sections of the society, but also in affluent households. Such deficiencies often lead to reduced vitality and fatigue and, if not tackled in time, may also lead to neurological, skeletal, or metabolic disorders (Di Somma et al. 2017; Christodoulou et al. 2013). This not only puts excess pressure on the state-funded healthcare systems, but also negatively affects individual efficiency and productivity as well as social structure. This problem is further accentuated by increasing medical costs. Genetic predisposition is certainly a crucial factor, but other factors, such as exposure to toxins, stress, and the food we eat, have significant impact on disease incidence and progression. Since the kind of food that we eat has an important influence on our health, it is vital that our food contains agents that may be helpful in preventing and/or treating diseases.

1.3 Cues from Traditional Systems of Medicine

The use of plants or their parts/extracts/concoctions occupies center stage in almost all the traditional systems of medicine (TSM), practiced worldwide. Fruits of the genus *Citrus* are generally rich in Vitamin C, folic acid, pectin, and potassium. They have been used in TSM to relieve palpitation, flatulence, as treatment for coughs and colds, as antispasmodic, antiemetic, antihypertensive, antipyretic, antiseptic, antihyperlipidemic, etc. (Turner and Burri 2013). Nutraceutical products based on bioflavonoids of *Citrus* are available for prevention as well as amelioration of cardiovascular problems. *Ginkgo biloba* has been traditionally used for treatment of memory impairment (Mullaicharam 2013). Several nutraceutical products based on this plant are available for treatment of dementia and improvement of memory (Chauhan et al. 2013). Similarly, green tea, which is popularly used as a nutraceutical, consumed as decoction or extract, finds its roots in traditional Chinese medicine. It was used for reduction of weight, positive effect on metabolism, prevention of heart ailments, etc. (Ahmad et al. 2014; Shinde et al. 2014). Garlic (*Allium cepa*) has been used in TSM worldwide, for various ailments such as fevers, swellings, and antibacterial (Chauhan et al. 2013). Garlic contains sulfur compounds: agoene, allicin, and alliin, minerals such as selenium, amino acids: cysteine, methionine, glutamine and isoleucine, flavonoids: cyanidin, allistatin I & II, quercetin, vitamins: A, B, C and E (Ayaz and Alpsoy 2007). Garlic finds use in prevention of cardiovascular diseases, regulation of blood pressure as well as reduction of cholesterol and blood sugar (Bhagyalakshmi et al. 2005; Wang et al. 2017). Garlic heart-care capsules are available as nutraceuticals. These are few popular examples where cues from TSM have been used for preparation of nutraceutical products. The use of herbs and phytochemicals in nutraceutical products will be revisited in the later section of this chapter, where it will be discussed in detail.

1.4 Contrasting TSM with Modern Medicine

Preventive health care in modern medicine is practiced at primary level by popularizing healthy living through regular exercise and appropriate diet (Finckh and Deane 2014), immunizations, and use of specific aids to protect from workplace hazards, such as use of earplugs in aerodromes. Secondary-level prevention is carried out by risk profiling through family history and epidemiological data, regular health check-ups to identify diseases at presymptomatic stage, and early treatment. Tertiary level includes more elaborate procedures such as use of cardiac stents for prevention of possible myocardial infarction, removal of tumor tissue to prevent metastasis, etc. (Karunathilake and Ganegoda 2018). On the other hand, disease prevention through dietary changes and/or inclusion of specific minerals, herbs, or herbal extracts in diet formed a major pillar of TSM (Pandey et al. 2013). Further, use of allopathic or modern medicine is popularly believed to be associated with side effects (Bandaranayake

2006), while the general public perception toward herbal drugs is positive and they are thought to be free from toxic side effects. Nutraceuticals draw similarity and strength from TSM, and the consumer perception toward them is quite positive, as evident from the rising market share of nutraceutical products (El Sohaimy 2012; Whitman 2001).

2 Major Classes of Nutraceutical Products

2.1 Herbal Extract/Plant-Based

Nutraceuticals hold an immense potential for health promotion and prevention of diseases using botanical products (Nasri et al. 2014). Compounds such as polyphenols, carotenoids, limonoids, flavonoids, and bioflavonoids obtained from different herbs like green tea, pomegranate, ginger, turmeric, and citrus fruits exhibit pharmaceutical properties and show promise as nutraceuticals (Shen et al. 2012). In the market, various nutraceutical products containing herbs are available in the form of capsules, powder, tablets, extract, syrup, etc., like chyawanprash, green tea, psyllium husk, brahmi, mucuna, etc. *Plantago ovata* (Plantaginaceae) is a source of Psyllium husk which is popularly used as a dietary supplement for its laxative properties. It is available in the form of granules, capsules, powder, or flakes. It has been found to be helpful in reducing cholesterol in addition to the treatment of constipation (Xing et al. 2017). It has also been shown to relieve diarrhea, improve appetite, and aid weight loss (Wärnberg et al. 2009; Mehmood et al. 2011; Verma and Mogra 2013). Brahmi or *Bacopa monnieri* (Scrophulariaceae) has been used as a tonic for mental health since ages (Gohil and Patel 2010). The key constituents of Brahmi are triterpenoid saponins: asiaticoside, madecassoside, bacosides A and B (Russo and Borrelli 2005; Kapoor 2017). Pharmacological studies have shown that Brahmi possesses activities such as memory enhancement (Kunte and Kuna 2013), antidepressant (Kadali et al. 2014), anti-Alzheimer's, neuroprotective, etc. (Rao et al. 2012). *Mucuna pruriens* (Fabaceae), popularly known as velvet-bean or dopa-bean, due to high content of L-dopa, is used for treatment of nervous disorders and Parkinson's disease (Patil et al. 2015). Several nutraceutical products based on *Mucuna* are available in the form of tablets, for treatment of stress, mood elevation, and mental well-being. *Tinospora cordifolia* (Menispermaceae), commonly called as giloy or guduchi, is used for treatment of fevers in traditional medicine (Kavya et al. 2015). It possesses anti-inflammatory and antipyretic properties and is often used in conditions like dengue, swine flu, malaria, and urinary tract infections (Saha and Ghosh 2012). Green tea is made from the leaves of *Camellia sinensis* (Theaceae) and is one of the most commonly consumed beverages around the world. It is known to harbor several health benefits such as cardiovascular protection, weight maintenance, skin care, allergy inhibition, and protection from osteoarthritis (Katiyar and Raman 2011; Wu et al. 2012; Zink and Traidl-Hoffmann 2015). Green tea has thermogenic and metabolic stimulant activity and is a rich source of antioxidants such as epigallocatechin 3-gallate (Nagle et al. 2006), vitamin C and E. Curcumin, the main bioactive compound present in the rhizome of *Curcuma longa* (Zingiberaceae), is well known for its anti-inflammatory properties (Kohli et al. 2005). Being fat-soluble, the bioavailability of curcumin is very low, and so in most ayurvedic formulations, it is taken along with black pepper, which contains piperine that acts as a bioavailability enhancer (Shoba et al. 1998). Curcumin finds use as a health supplement to relieve inflammation and arthritis (Hewlings and Kalman 2017). Ginger, the rhizome of *Zingiber officinale* (Zingiberaceae), is a widely used condiment in Asian cuisines (Sharma 2017). Capsules/tablets made from ginger are used as nutraceutical for their antiemetic, carminative, and anti-inflammatory properties (Malhotra and Singh 2003; Palatty et al. 2013; Funk et al. 2016). Ashwagandha, the roots of *Withania somnifera* (Solanaceae), are known in traditional medicine for their restorative and rejuvenating benefits (Sharma et al. 2011). It is widely used as a nutraceutical in various forms like powder, capsules, tablets, etc., for providing benefits such as vigor enhancement, and stress management (Lopresti et al. 2019). *Emblica officinalis* (Phyllanthaceae), known as Indian gooseberry or amla, is an ingredient of Triphala, one of the popular nutraceuticals consumed in India (Peterson et al. 2017). Triphala is used for its properties in aiding digestion, relieving constipation and anti-inflammatory action (Mukherjee et al. 2006; Kalaiselvan and Rasool 2015). *Azadirachta indica* (Meliaceae), popularly known as neem, is an important medicinal herb known to purify blood, detoxify body, and neutralize free radicals. It also supports immune system, provides radiant skin, supports healthy digestion, boosts liver function, etc., and is available in the form of tablets and capsules (Mandal-Ghosh et al. 2007; Bhowmik et al. 2010; Pingale Shirish 2010). *Ginkgo biloba* (Ginkgoaceae) has been used in traditional Chinese medicine for the treatment of respiratory diseases, cardiovascular diseases, neurodegenerative diseases such as Alzheimer's disease and fatigue for several years (Ramamurthy et al. 2014; Mahadevan and Park 2008). Ministry of Food and Drug Safety (MFDS) of Korea has accepted the use of *Ginkgo biloba* for memory enhancement and improving blood circulation. EGb 761® extract obtained from the leaves of *Ginkgo biloba* has displayed neuroprotective and cardiovascular protective property (Maclennan et al. 2002; Wąsik and Antkiewicz-Michaluk 2017; McKeage and Lyseng-Williamson 2018) and has been shown to ameliorate mild memory and cognitive impairment.

2.2 Proteins and Amino Acids

Proteins are an important component of normal diet and are a must requirement for healthy functioning of human body. They are important structural and functional units in all life forms. A large population, especially in developing countries, is known to consume diets deficient in proteins (Müller and Krawinkel 2005). Traditional vegetarian diets, in general, are low in protein content. Further, protein-digestibility corrected amino acid score (PDCAAS) of pulses is generally much lower compared to meat and egg. Milk and soy, though have a good PDCAAS (Schaafsma 2000), may not suit the diets of all individuals due to issues such as lactose intolerance and casein allergy (milk) and adverse effect on thyroid function (soy). However, to address these issues, several nutraceutical products in the form of protein powders, cookies, shakes, etc., are available in the market. Many of these products may contain total whey or soy protein, or their hydrolysates (Manninen 2009). Further, obese people who intend to lose weight are advised low-carbohydrate–low-fat–high-protein diets. Such individuals also consume protein shakes. Athletes, who require higher muscle mass, often supplement their diets with protein-based nutraceuticals (Phillips and Van Loon 2011). Many such products also contain higher amounts of specific amino acids such as L-arginine. L-arginine is metabolized to nitric oxide, which increases blood flow, thereby helping in better muscle development (Álvares et al. 2012). L-arginine is also metabolized to creatinine that facilitates better muscle mass and higher anaerobic work capacity. L-arginine also helps to increase the levels of resting growth hormone, which in turns helps in gaining muscle mass (Kanaley 2008). L-Arginine soft-gel helps in reduction of menopause symptoms (Stanislavov and Rohdewald 2014). Agmatine powder has been used to enhance rate of absorption and overall bioavailability (Schwedhelm et al. 2008). Some of these supplements may also contain branched chain amino acids (BCAA) like leucine, valine, and isoleucine which are metabolized into glutamine and alanine (Holeček 2018). These help to increase immunity, boost muscle growth, provide neuro-protection, and benefit people with liver disease (Shimomura et al. 2006; Fernstrom 2005; Tajiri and Shimizu 2013). Aromatic acid tryptophan is also available as supplement for helping thyroid hormone biosynthesis and maintaining metabolism. Similarly, phenylalanine and tryptophan supplementation is known to provide relief from stress and anxiety and improve mental health (Fernstrom and Fernstrom 2007; Jenkins et al. 2016). Shark cartilage powder that mainly contains the collagenous protein which forms the tough skeleton of shark is also available as a nutraceutical. It is thought to support and repair joint tissues as well as for the prevention and/or control of arthritis (Merly and Smith 2015).

2.3 *Fatty Acids*

Polyunsaturated fatty acids (PUFAs) contain at least one double bond in their backbone and may be either omega-3 (n-3) or omega-6 (n-6) fatty acids. α-linolenic acid (ALA), eicosapentaenoic acid (EPA), and docosahexaenoic acid (DHA) are examples of omega-3 FA. They are crucial for normal functioning of the body. Various studies suggest that omega-3-fatty acids have anti-arrhythmic (Leray et al. 2001; Stoll et al. 1999), hypolipidemic (Bucher et al. 2002; Nemets et al. 2002), and antithrombotic (Stoll et al. 1999; Bucher et al. 2002; Albert et al. 2002) activities. Further, they are important for brain development as well as its proper functioning. Supplementing diet with omega-3 (EPA + DHA) fatty acids has also been shown to boost immunity, cardiovascular health, and cognition ability and provide relief from joint pain. These have to be acquired through the diet. However, mostly vegetarian food is deficient in both EPA and DHA (Escott-Stump and Mahan 2000); hence, dietary supplements containing these are popular. Omega-3 (ALA, EPA, and DHA) are present in several dietary supplement formulations, such as fish oil, krill oil, and cod liver oil soft-gel capsules (Fialkow 2016). Omega-6 FA mainly includes linoleic acid (LA), γ-linolenic acid (GLA), and arachidonic acid (ARA). The main source of LA is vegetable oils such as corn, safflower, soybean, and sunflower. ARA is mainly obtained from animal products like meat, poultry, and eggs. Some observational studies predict that the alpha-linolenic acid is a promising nutraceutical for the prevention of stroke (Blondeau et al. 2015).

2.4 *Small Molecules, Vitamins, and Minerals*

Glucosamine and chondroitin supplementation has been reported to improve joint pain, swelling, and physical function of joints. Nutraceutical products comprising both glucosamine and chondroitin have been found to be helpful to osteoarthritis patients (Huskisson 2008; Vasiliadis and Tsikopoulos 2017). Similarly, Vitamin D and calcium are important for maintenance of bone health. Due to changing lifestyle and eating habits as well as relatively reduced exposure to sun, vitamin D deficiency is becoming commonplace. Several supplements containing vitamin D, calcium along with glucosamine and chondroitin are available to help skeletal function (Jerosch 2011; Abrams et al. 2018). Vitamin D deficiency has also been related to poor immune response to infections and increased incidence of autoimmune disorders and has been associated with neuropsychiatric disorders (Holick 2004; Sotodeh-Asl et al. 2014). Similarly, deficiency of vitamin B12 is associated with tiredness, weakness, shortness of breath, pale skin, as well as nerve problems like numbness or tingling. Vegetarian diet is perpetrated to be deficient in vitamin B12 (Antony 2003). Several multivitamin dietary supplements are available to address this need. Dietary supplements with combinations of vitamins with whey protein lysate or with omega-3 fatty acids are also available in the market (Andreeva et al. 2012; Kemse et al. 2014; Chanet et al. 2017). A large population in developing countries also suffers from anemia, due to deficiency of vitamin B12 and iron. Several supplements are available which club iron and multivitamin to address this need. Due to increased usage of reverse osmosis purified and demineralized water, the incidence of mineral deficiency, such as those of calcium, magnesium, potassium, zinc, selenium, and iron, is increasing. Deficiencies of minerals are associated with various diseases and disorders (Kozisek 2005). Some of these minerals are important for bone health, while others may be required for other physiological functions (Campbell 1995; New 1999). Several protein supplements available in the market include the recommended daily dose of the required vitamins and minerals, to address these needs.

2.5 Probiotics and Prebiotics

A probiotic comprises of alive microorganisms as dietary supplements, which when taken in an adequate amount benefits the health by improving intestinal microbial balance (Fuller 2012). It is mandatory that the microorganism to be used as a probiotics comes under the category of GRAS (generally recognized as safe) (Bouchard et al. 2013). The most common sources of probiotics are yogurt, cultured buttermilk, and cheese. Most probiotic preparations currently in the market consist of lactic acid bacteria (like *Lactobacilli, Streptococci*, and *Bifidobacteria*), which are predominately present in healthy human gastrointestinal microbiota (Das et al. 2012). They are readily available in the forms of powder, gel or paste or granule, capsule, etc. (Suvarna and Boby 2005). Various nutraceutical capsules consisting of *Lactobacillus rhamnosus, Lactobacillus plantarum, Lactobacillus salivarius*, etc., are known to boost immunity and maintain healthy intestinal flora which are readily available in the market. There are evidences that consuming probiotic supplements reduces the risk of systemic conditions, like allergy, asthma, urinary tract infections, and otitis in children (Lenoir-Wijnkoop et al. 2007). Many probiotics are mainly used to treat various gastrointestinal (GI) problems such as lactose intolerance, diarrhea and antibiotic-associated GI side effects (Doron et al. 2005), irritable bowel syndrome (IBS) (Lacy and De Lee 2005), etc. *Helicobacter pylori* is the causal agent of peptic ulcer disease and is also associated with carcinogenesis. It has been found that, in vitro, multiple strains of *Lactobacilli* and some flavonoids have the potential to reduce the *H. pylori* growth. Probiotics have also been shown to inhibit gastric mucosal adhesion of *H. pylori* (Ushiyama et al. 2003; Chatterjee et al. 2003). In addition to this, *Lactobacillus casei*-supplemented milk product to a standard triple therapy regimen (omeprazole, amoxicillin, and clarithromycin) in children resulted in a remarkably higher eradication rate of *H. pylori* and conferred an enhanced therapeutic benefit (Sýkora et al. 2005). Lactose intolerance is a genetically determined beta-galactosidase deficiency which results in the inability to break down lactose into the monosaccharides glucose and galactose. As a result, individuals suffering from lactose intolerance develop diarrhea, bloating, abdominal pain, flatulence, etc., after consumption of dairy products. It has been suggested that *Streptococcus thermophilus* and *Lactobacillus delbrueckii* subsp. *bulgaricus* have the ability to digest milk products due to the presence of beta-galactosidase enzyme (Kechagia et al. 2013).

Prebiotics are dietary ingredients which modulate gut microbiota composition and thereby may enhance metabolic rate, increase uptake of minerals, augment immunity level, and improve role of intestinal mucosal barrier (Nagpal and Kaur 2011; Anadón et al. 2016). The most widely accepted prebiotics are fructo-oligosaccharides (fruits and vegetables), inulin (Chicory root, *Agave tequilana*), and galacto-oligosaccharides (milk). Prebiotics may also include polydextrose, soybean oligosaccharides, isomalto-oligosaccharides, xylo-oligosaccharides, palatinose, dextrins, gentio-oligosaccharides, chito-oligosaccharides, and sugar alcohols (Anadón et al. 2016). Most prebiotics are short-chain carbohydrates with a degree of polymerization of two or more and are mainly not susceptible to digestion by pancreatic and intestinal brush border enzymes (Steed and Macfarlane 2009). Prebiotics can also be used as a supplement that can be taken either directly in the form of capsules and tablets or indirectly by sprinkling directly on food as well as by stirring into beverages. Lactulose, a non-absorbably sugar, is available as a prebiotic for relieving constipation and symptoms of hepatic encephalopathy (Gibson 2004; Conway 2001; Delzenne 2003). Similarly, inulin is helpful in alleviating inflammatory bowel disease by altering the intestinal microflora (Schultz et al. 2004; Cherbut et al. 2003; Furrie et al. 2005; Kelly et al. 2005).

2.6 Microalgae as Nutraceuticals

Microalgae like *Nostoc*, *Botryococcus*, *Anabaena, Chlamydomonas, Scenedesmus, Chlorella*, etc., contain minerals, vitamins, essential amino acids, fatty acids, and secondary metabolites, thus making it possible to exploit these as nutraceuticals with some of them possessing antioxidant and anti-aging properties (Bishop and Zubeck 2012). Microalgae-based nutraceuticals are becoming popular and are acquiring a large market share (Nicoletti 2016). Cell extracts or dried biomass produced from Spirulina or Chlorella is among the most common blue-green algae-based nutraceuticals (Suganya et al. 2016). They have been found to have positive effects on cardiovascular diseases, diabetes, hypertension, and blood pressure lowering and reducing cholesterol (Iwata et al. 1990). Spirulina is available in different forms such as tablets, capsules, or powder, alone or in combination with other algae, plant extracts, or vitamins, for both human and animal food. Iron-enriched Spirulina tablets are available as a food supplement to compensate iron requirement, reduce fatigue, and improve anemia. Chlorella contains high content of proteins, carotenoids, and vitamins and is used as a dietary protein supplement (Andrade et al. 2018). Dunaliella (*D. salina, D. bardawil*) is a unicellular green alga, known for the production of large amounts of carotenoid (β carotene), glycerine, and protein (Bental et al. 1988; Murthy et al. 2005). The algal products in various forms like β-carotene capsules, extracts, Dunaliella powder, and dried Dunaliella are accessible in the market for human and animals. Haematococcus (*H. pluvialis*) is a unicellular green alga that is used in several nutraceutical products mainly because of high content of astaxanthin along with other vitamins and minerals (Lorenz and Cysewski 2000). Astaxanthin possesses strong free-radical scavenging activity, which is 1000 times more that of vitamin E (α-tocopherol) (Naguib 2000). Astaxanthin has been shown to exhibit protection against diseases such as inflammation, antibacterial, diabetes, diabetic nerve damage, neurodegenerative, and eye diseases (Yuan et al. 2011). The annual world market of astaxanthin is estimated to be about $200 million (Irianto and Austin 2002;

Guerin et al. 2003). Aphanizomenon (blue-green alga) found in freshwater systems throughout the world is also used as a nutraceutical products (Pulz and Gross 2004) as it produces omega-3 and omega-6 polyunsaturated fatty acids in large quantity (DeWille et al. 1979). Figure 1 summarizes the major classes, and Table 1 lists a few examples of nutraceutical products available in the market.

3 Regulatory Aspects of Nutraceuticals

It is an important to regulate nutraceutical products to protect the public from spurious and possibly toxic products (Bagchi et al. 2004; Halsted 2003; Hasler 2002). As previously mentioned, a class of nutraceutical products includes natural products directly or as a component. The natural products industry faces diverse challenges starting from the collection of raw material to manufacturing. Collection of sample, identification, quantification, and standardization methods are of critical importance for maintenance of uniform quality. Parameters asserting product safety and efficacy, such as acute and chronic and toxicity studies, genotoxicity, reproductive toxicology, teratogenicity, etc., should be carried out. If long term consumption of nutraceutical is intended, then supplementation studies in animals and limited clinical trials may also be necessary (Bagchi 2006). Regulations should require that the nutraceuticals be judged safe before they are marketed, especially considering that they may be taken over long periods of time and would be available over-the-counter without the need for prescription. Nutraceuticals are predominately considered as food product; however, their classification and regulation may vary in different regions of the world. So, they are regulated differently in different jurisdictions (Table 1).

UNFAO (United Nation food and Agricultural Organizations) and WHO (World Health organization) have given specific guidelines for the production of and marketing foodstuffs and their derivatives. Different countries employ these guidelines for the regulation of dietary supplements, which briefly explain health claims according to the nutrient function; enhanced function; and reduction of probability of diseases. The nutrient function claim can be defined as the correlation between the role of nutrient and function of the human body. In the USA, Food and Drug Administration (USFDA) defines nutraceuticals or dietary supplements as products other than tobacco that may be as tablet, capsule, soft-gel, or powder which is intended to supplement the diet or enhance the health. Such a product may contain vitamins, minerals, amino acids, botanicals, or other dietary substance (Hoadley and Rowlands 2014). In USA, the FDA is also responsible for taking legal action against any adulterated supplement product after it reaches into the market (Santini et al. 2018). Under the Dietary Supplement Health and Education Act (DSHEA 1994), it is the liability of manufacturers and distributors of dietary supplements for the evaluation of the safety, efficacy, and labeling of all the items before marketing to ensure that the safety of the nutraceutical. The USFDA Modernization Act (1997) mentions that at least four months before a supplement is marketed, the FDA has to be notified about the health claims and/or the nutrient content claims on the product label of a dietary

Table 1 Nutraceuticals along with their source and applications

S. No	Category	Species/product	Application	References
1	Herb	*Camellia sinensis* (Epigallocatechin gallate)	Free radical scavenger Antioxidant	Nagle et al. (2006)
2	Herb	*Curcuma longa* (Curcuminoids)	Antioxidant Anti-inflammatory	Hewlings and Kalman (2017)
3	Herb	*Mucuna pruriens* (L-Dopa)	Enhance dopamine levels and protects neurons, promote brain health	Patil et al. (2015)
4	Plant	*Glycyrrhiza glabra* (liquorice)	Free radical scavenger, Hepato-protective	Kalsi et al. (2016)
5	Herb	*Panax ginseng* (Ginsenoside)	Anti-inflammatory	Kim et al. (2017)
6	Herb	*Ocimum sanctum* (Eugenol)	Anti-microbial Anti-stress Anti-diabetic Hepatoprotective Anti-inflammatory Neuroprotective Cardioprotective	Baliga et al. (2013), Prakash and Gupta (2005)
7	Plant	*Scutellaria baicalensis* (Baicalein)	Antioxidant Anti-inflammatory	Shieh et al. (2000)
8	Plant	*Azadirachta indica*	Immune system, radiant skin, support healthy digestion, boosts liver function	Mandal-Ghosh et al. (2007), Bhowmik et al. (2010), Pingale Shirish (2010)
9	Plant	*Coffea arabica* Caffeine	Regulate oxidative stress	Martini et al. (2016)
10	Herbal extract	*Ginkgo biloba* extract (EGb 761)	Memory enhancement, Improvement in blood circulation	Maclennan et al. (2002), Wąsik and Antkiewicz-Michaluk (2017), McKeage and Lyseng-Williamson (2018)
11	Mineral	Magnesium	Required for the active transport of ions like K and Ca and provide strength to the bone mass	Bhutto et al. (2005)
12	Herb	*Larrea tridentata* (Nordihydroguaiaretic acid)	Antioxidant Antiviral Anti-inflammatory	Rahman et al. (2011)

(continued)

Table 1 (continued)

S. No	Category	Species/product	Application	References
13	Mineral	Vitamin B complex	Neuro modulatory function, bone strength	Kennedy (2016), Dai and Koh (2015)
14	Vitamins	Vitamin C and E	Free radical scavenger	Grosso et al. (2013)
15	Vitamin	Vitamin D and Vitamin C	Maintain calcium homoeostasis, aging Immuno-modulatory effects Regulate dopamine levels	Veldurthy et al. (2016), Harrison and May (2009)
16	Mineral	Calcium supplement	Strength to the bone	Reid et al. (2015)
17	Amino-acids	L-Arginine softgel	Reduction of menopause symptoms	Stanislavov and Rohdewald (2014)
18	Amino-acids	Agmatine powder	Enhance rate of absorption	Schwedhelm et al. (2008)
19	Amino acids	BCAA (Branched Chain Amino-Acids)	Increase immunity, provide neuro-protection as well as benefit people with liver disease	Shimomura et al. (2006), Fernstrom (2005), Tajiri and Shimizu (2013)
20	Amino-acids	Aromatic amino acids	Relief from stress, anxiety and improve mental health	Fernstrom and Fernstrom (2007)
21	Herbs	*Centella asiatica*/Brahmi	Nerve tonic, anti-anxiety, spasmolytic	Chauhan et al. (2013)
22	Herbs	*Aegle marmelos*/Bael	Digestive, appetizer, treatment of diarrhoea and dysentery	Neeraj and Johar (2017)
23	Plant	*Ferula assafoetida* (Asafoetida/ferulic and umbellic acid)	Stimulant, expectorant, carminative, laxative, etc	Mahendra and Bisht (2012)
24	Herb	*Panax ginseng* (Ginsenosides and Panaxosides)	Stimulate immune and nervous system	Kang and Min (2012)
25	Herb	Ginger (Zingiberene and gingerols)	Stimulant, chronic bronchitis, hyperglycemia	Sharma (2017)

(continued)

Table 1 (continued)

S. No	Category	Species/product	Application	References
26	Herb	*Echinacea purpurea* (alkylamide and echinacoside)	Anti-inflammatory, anti-viral and immunomodulatory	Manayi et al. (2015)
27	Plant	*Allium sativum* (Alliin and Allicin)	Anti-inflammatory, immunomodulation, nervine tonic	Arreola et al. (2015)
28	Herb	*Tinospora cordifolia* Giloy	Anti-inflammatory and anti-pyretic activity	Saha and Ghosh (2012)
29	Herb	*Withania somnifera* (Ashwagandha)	Aphrodisiac, liver tonic, anti-inflammatory agent, astringent, asthma	Sharma et al. (2011)
30	Herb	*Cassia angustifolia* (Sennosides)	Purgative	Ramchander and Middha (2017)
31	Herb	*Emblica officinalis* (Amla)	Anti-inflammatory, fever, anaemia, etc	Jain et al. (2015)
32	Herb	*Hydrastis Canadensis* (Hydrastine and berberine)	Anti-microbial, astringent, treatment of mucosal inflammation	Asmi and Lakshmi (2013)
33	Herb	*Valeriana officinalis* (valerenic acid and valerate)	Tranquillizer, migraine, intestinal cramps, bronchial spasm	Pilerood and Prakash (2013)
34	Minerals	Glucosamine and chondroitin	Treatment of osteoarthritis	Huskisson (2008), Vasiliadis and Tsikopoulos (2017)
35	Herb	*Glycyrrhiza glabra* Liquorice	Anti-inflammatory and anti-allergy	Shin et al. (2007)
36	Prebiotic	Lactulose	Treatment of liver diseases and relief from symptoms of constipation	Gibson (2004), Conway (2001), Delzenne (2003), Marteau and Boutron-Ruault (2002)
37	Prebiotic	Fructo-oligosaccharides	Treatment of inflammatory bowel disease and relief from constipation	Cherbut et al. (2003), Schultz et al. (2004), Furrie et al. (2005), Kelly et al. (2005)

(continued)

Table 1 (continued)

S. No	Category	Species/product	Application	References
38	Prebiotic	Raffinose, galacto-oligosaccharide, isomalto-oligosacharides	Prevention of cholesterol gallstones	Mitsuoka et al. (1987), Kohmoto et al. (1988)
39	Microalgae	*Arthrospira platensis*	Shows anti-oxidant and antibacterial property	Nuhu (2013)
40	Microalgae	*Chlorella vulgaris*	Health food, food Supplement, feeds	Batista et al. (2013), Paniagua-Michel (2015)
41	Microalgae	*Dunaliella salina* (Lutein, betacarotene)	Free radical scavenger and provide hepato-protection	Murthy et al. (2005), Hsu et al. (2008)
42	Microalgae	*Haematococcus pluvialis* (Astaxanthin)	Health food, feeds	Paniagua-Michel (2015)
43	Microalgae	*Schizochytrium* (Docohexaenoic acid)	Dietary, nutritional supplements	Chu (2012)
44	Micro algae	*Aphanizomenon flos-aquae* Mycosporine-like amino acids	UV-screening agent; sunscreen	Chu (2012)
45	Micro algae	*Crypthecodinium cohnii*	Oil for the infant formula as DHA source	Saha and Murray (2018)

supplement which may be authorized by a statement of the Academy of Sciences or another federal body. European Food Safety Authority (EFSA) approves the health claim of a nutraceutical product before it can be put on the market. Over and above the EFSA's opinion on a product, the member states can individually set specific regulations for approval. However, unlike the USFDA, EFSA is not mandated to take a legal action against an unsafe product (Hasler 2005). In Canada, the regulation of nutraceuticals is more like a drug than food category (L'Abbé et al. 2008). In India, these products are regulated by the Food Safety and Standards Regulations (2016). These regulations cover nutraceuticals, food for special dietary requirement or for special medical conditions. These products may include botanicals, prebiotics, probiotics, etc. However, these regulations do not allow the use of hormones, steroids, or psychotropic substances. FSSAI has published a list of dietary components and their maximum daily doses. As such, there are no requirements for any non-clinical or clinical data as long as the product contains a dietary component that is listed by the FSSAI and follows other regulations regarding permitted dosage and labeling norms. In case a new substance is to be added to a product, then safe history of human usage, scientific statements by academia, acute or chronic safety and toxicology studies, and limited human trials may be required. Similarly, in Japan FoSHU (Food for Specific Health Uses) system was introduced in 1991 by the Ministry of Health and Welfare, now known as the Ministry of Health, Labor, and Welfare (MHLW), as a regulatory system to approve statements concerning the effects of the food on the human body (Shimizu 2003). So, for the approval of food supplements with health-promoting activities, only requirements by FoSHU are safety of food, nutritional value of food ingredients, etc., even if these perpetrated pharmacological activities are not validated with any scientific proof (Santini et al. 2018; Saito 2007). In any countries, like in Australia or China, the nutraceuticals are regulated mainly as a category of food (Tee et al. 2002; Tapsell 2008; Yang 2008). Detailed evaluation of nutraceutical product with clinical trial as well as safety assessment study should be carried out before it is marketed. A health claim substantiated with safety and efficacy data, based on an understanding of the mode of action and the absence of any unwanted side effects, further validated through clinical evidence should ideally be required, in the interest of public safety, for a nutraceutical product to be approved.

4 Business Outlook

Due to the increasing prevalence of lifestyle diseases and rising awareness regarding preventive healthcare measures, the demand of nutraceutical has soared worldwide, during the past two decades. Initially, from 1999 to 2002 the nutraceutical industry grew at 7% per year, the next few years up to 2010 saw almost twice that growth at 14% per annum (Verma and Popli 2018). In 2017, the nutraceutical market is estimated to be worth $379 billion. It has been predicted that globally the nutraceutical market was $230.9 billion in 2018 which is expected to reach $336.1 billion by 2023 at a compound annual growth rate (CAGR) of 7.8%, according to the BCC research report Nutraceuticals: Global Market to 2023 (www.bccresearch.com/market-research/food…nutraceuticals-global-markets). Consumption of dietary supplements is expected to rise at a CAGR of over 9.7% from 2017 to 2025 worldwide due to the increasing awareness about preventive health care. These supplements are available in various forms like herbal extracts, tablets, capsules, powders, etc. Their low-cost, easy over-the-counter accessibility, and health-promoting properties are the main driving factors anticipated to increase their demand over the next few years (www.grandviewresearch.com/industry-analysis/nutraceuticals-market). In Europe also the nutraceutical market is dominated by the dietary supplement with the market share of 30.1% in 2016 and is predicted to grow at a CAGR of 6.4% from 2017 to 2025. The dietary supplements market, comprising vitamins, are the most common supplement consumed by the people and account for 46.8% out of all total dietary supplements followed by the herbal extract with an estimated growth at a CAGR of 9.8%. The nutraceuticals comprising proteins and amino acids may grow at a CAGR 9.4% from 2017 to 2025 (www.figlobal.com). In Africa, the infant nutrition industry and the demand for nutritious food and beverages are the key drivers for the nutraceutical market's growth. Countries, like Egypt and South Africa, are trying to acquire more herb-based nutraceutical due to the larger acceptability. This may offer good market opportunity in the Africa. However, the microalgae-based omega-3-fatty acid is the new emerging field in the nutraceutical market of Africa (www.mordorintelligence.com). Vitamins and minerals dominated the Chinese market followed by the herbal supplements segment. According to the report Indian Nutraceuticals Market Outlook: Vision 2022 published by ASSOCHAM and RNCOS, the Indian dietary supplement market was valued at US$2.8 billion in 2015 which is expected to reach a value of US$8.5 billion by 2022. The nutraceutical market for herbal supplements in India was valued at approximately US$0.6 billion in 2015, which is expected to reach a value of US$1.7 billion by 2022. The vitamin and mineral market in India is expected to grow in the coming years and estimated to reach at a value of US$2.1 billion by 2022. Similarly, proteins and amino acids are an important class of supplements that have gained a lot of popularity in the recent years. From 2015 to 2022, the market for these supplements is expected to rise from US$0.4 billion to US$1.09 billion by 2022. The market for the dietary supplements including fatty acids and antioxidants in India was valued at approximately US$0.1 billion in 2015 and is estimated to reach a value of US$0.23 billion by 2022. In 2015, the market for functional foods like oats, soy, probiotic yogurt, fortified baked goods, and fortified edible oils was valued at US$0.7 billion and expected to rise at worth of US$2 billion by 2022. Similarly, the market for functional beverages was US$0.3 billion in 2015 and predicted to reach an approximate value of US$1.1 billion by 2022 (www.assocham.org/newsdetail.php?id=6259).

Thus, in conclusion, it may be summarized that increasing awareness, positive public perception, and ever-increasing medical costs may push the market of nutraceuticals which are perceived as one of the major drivers of preventive health care. The market and hence business opportunities for nutraceutical products may continuously show an upward trend for next few years. Indeed in countries such as India, a lot of start-up countries are focusing on this segment. However, with such a high growth, there is a need for strong regulatory setup so as to protect the interest of consumers.

References

Abrams GD, Feldman D, Safran MR (2018) Effects of vitamin D on skeletal muscle and athletic performance. J Am Acad Orthop Sur 26(8):278–285

Ahmad M, Baba WN, Shah U, Gani A, Gani A, Masoodi FA (2014) Nutraceutical properties of the green tea polyphenols. J Food Process Technol 5(11):1–5

Albert CM, Campos H, Stampfer MJ, Ridker PM, Manson JE, Willett WC, Ma JJ (2002) Blood levels of long-chain n–3 fatty acids and the risk of sudden death New England. J Med 346(15):1113–1118

Álvares TS, Conte CA Jr, Paschoalin VMF, Silva JT, Meirelles CdM, Bhambhani YN, Gomes PS (2012) Acute l-arginine supplementation increases muscle blood volume but not strength performance. Appl Physiol Nutr Metab 37(1):115–126

Anadón A, Martínez-Larrañaga MR, Ares I, Martínez MA (2016) Prebiotics: safety and toxicity considerations. In: Nutraceuticals. Academic Press, Cambridge

Andrade LM, Andrade CJ, Dias M, Nascimento C, Mendes MA (2018) Chlorella and *Spirulina* Microalgae as Sources of Functional Foods, Nutraceuticals and Food Supplements. MOJ Food Process Technol 6(2):1–15

Andreeva VA, Galan P, Torres M, Julia C, Hercberg S, Kesse-Guyot E (2012) Supplementation with B vitamins or n−3 fatty acids and depressive symptoms in cardiovascular disease survivors: ancillary findings from the SUpplementation with FOLate, vitamins B-6 and B-12 and/or OMega-3 fatty acids (SU. FOL. OM3) randomized trial. Am J Clin Nutr 96(1):208–214

Antony AC (2003) Vegetarianism and vitamin 8-12 (cobalamin) deficiency. Am J Clin Nutr 78(1):3-6

Arreola R, Quintero-Fabián S, López-Roa RI, Flores-Gutiérrez EO, Reyes-Grajeda JP, Carrera-Quintanar L, Ortuño-Sahagún D (2015) Immunomodulation and anti-inflammatory effects of garlic compounds. J Immunol Res 2015:1–13

Asmi S, Lakshmi T (2013) Therapeutic aspects of goldenseal. IRJP 4(9):41–43

Ayaz E, Alpsoy H (2007) Garlic (*Allium sativum*) and traditional medicine. Turkiye Parazitol Derg 31(2):145–149

Bagchi D (2006) Nutraceuticals and functional foods regulations in the United States and around the world. Toxicology 221(1):1–3

Bagchi D, Preuss HG, Kehrer JP (2004) Nutraceutical and functional food industries: aspects on safety and regulatory requirements. Toxicol Lett 1(150):1–2

Bahadoran Z, Mirmiran P, Azizi F (2015) Fast food pattern and cardiometabolic disorders: a review of current studies. Health Promot Perspect 5(4):231–240

Baliga MS, Jimmy R, Thilakchand KR, Sunitha V, Bhat NR, Saldanha E, Rao S, Rao P, Arora R, Palatty PL (2013) *Ocimum sanctum* L (Holy Basil or Tulsi) and its phytochemicals in the prevention and treatment of cancer. Nutr Cancer 65(Supp 1):26–35

Bandaranayake WM (2006) Quality control, screening, toxicity, and regulation of herbal drugs. Turning Medicinal Plants into Drug. Wiley-Vch, Weinheim, Modern Phytomedicine

Batista AP, Gouveia L, Bandarra NM, Franco JM, Raymundo A (2013) Comparison of microalgal biomass profiles as novel functional ingredient for food products. Algal Res 2(2):164–173

Bental M, Degani H, Avron M (1988) 23Na-NMR studies of the intracellular sodium ion concentration in the halotolerant alga *Dunaliella salina*. Plant Physiol 87(4):813–817

Bhagyalakshmi N, Thimmaraju R, Venkatachalam L, Murthy KC, Sreedhar R (2005) Nutraceutical applications of garlic and the intervention of biotechnology. *Crit Rev Food Sci Nutr 45*(7–8):607–621

Bhowmik D, Chiranjib YJ, Tripathi K, Kumar KS (2010) Herbal remedies of *Azadirachta indica* and its medicinal application. J Chem Pharm Res 2(1):62–72

Bhutto A, Mastoi AA, Memon SA, Qureshi GA, Qureshi AA (2005) Magnesium and its essential role in health. JLUMHS 4(1):33–35

Bishop W, Zubeck HJJNFS (2012) Evaluation of microalgae for use as nutraceuticals and nutritional supplements. J Nutr Food Sci 2(5):1–6

Blondeau N, Lipsky RH, Bourourou M, Duncan MW, Gorelick PB, Marini AM (2015) Alpha-linolenic acid: an omega-3 fatty acid with neuroprotective properties—Ready for use in the stroke clinic? BioMed Res Int 2015:1–8

Bouchard DS, Rault L, Berkova N, Le Loir Y, Even S (2013) Inhibition of *Staphylococcus aureus* invasion into bovine mammary epithelial cells by contact with live *Lactobacillus casei*. Appl Environ Microbiol 79(3):877–885

Bucher HC, Hengstler P, Schindler C, Meier G (2002) N-3 polyunsaturated fatty acids in coronary heart disease: a meta-analysis of randomized controlled trials. Am J Med 112(4):298–304

Campbell JD (1995) Minerals and disease. J Orthomol Med 10:177–188

Cencic A, Chingwaru W (2010) The role of functional foods, nutraceuticals, and food supplements in intestinal health. Nutrients 2(6):611–625

Chanet A, Verlaan S, Salles J, Giraudet C, Patrac V, Pidou V, Pouyet C, Hafnaoui N, Blot A, Cano N (2017) Supplementing breakfast with a vitamin D and leucine–enriched whey protein medical

nutrition drink enhances postprandial muscle protein synthesis and muscle mass in healthy older men. J Nutr 147(12):2262–2271
Chatterjee A, Yasmin T, Bagchi D, Stohs SJ (2003) The bactericidal effects of *Lactobacillus acidophilus*, garcinol and Protykin® compared to clarithromycin, on *Helicobacter pylori*. Mol Cell Biochem 243(1–2):29–35
Chauhan B, Kumar G, Kalam N, Ansari SH (2013) Current concepts and prospects of herbal nutraceutical: A review. J Adv Pharm Technol Res 4(1):4–8
Cherbut C, Michel C, Lecannu G (2003) The prebiotic characteristics of fructooligosaccharides are necessary for reduction of TNBS-induced colitis in rats. J Nutr 133(1):21–27
Christodoulou S, Goula T, Ververidis A, Drosos G (2013) Vitamin D and bone disease. Biomed Res Int 2013:1–6
Chu WL (2012) Biotechnological applications of microalgae. IeJSME 6(1):S24–S37
Conway PLJN (2001) Prebiotics and human health: the state-of-the-art and future perspectives. Näringsforskning 45(1):13–21
Coussens LM, Werb Z (2002) Inflammation and cancer. Nature 420(6917):860–867
Dai Z, Koh WP (2015) B-vitamins and bone health–a review of the current evidence. Nutrients 7(5):3322–3346
Das L, Bhaumik E, Raychaudhuri U, Chakraborty R (2012) Role of nutraceuticals in human health. J Food Sci Technol 49(2):173–183
DeFelice SL (1995) The nutraceutical revolution: its impact on food industry R&D. Trends Food Sci Technol 6(2):59–61
Delzenne NM (2003) Oligosaccharides: state of the art. P Nutr Soc 62(1):177–82
DeWille JW, Fraker PJ, Romsos DR (1979) Effects of essential fatty acid deficiency, and various levels of dietary polyunsaturated fatty acids, on humoral immunity in mice. J Nutr 109(6):1018–1027
Di Somma C, Scarano E, Barrea L, Zhukouskaya VV, Savastano S, Mele C, Scacchi M, Aimaretti G, Colao A, Marzullo P (2017) Vitamin D and neurological diseases: an endocrine view. Int J Mol Sci 18(11):1–26
Doron S, Snydman DR, Gorbach SL (2005) *Lactobacillus* GG: bacteriology and clinical applications. Gastroenterol Clin North Am 34(3):483–498
Dureja H, Kaushik D, Kumar V (2003) Developments in nutraceuticals. Indian J Pharmacol 35(6):363–372
El Sohaimy S (2012) Functional foods and nutraceuticals-modern approach to food science. World Appl Sci J 20(5):691–708
Elorinne AL, Alfthan G, Erlund I, Kivimäki H, Paju A, Salminen I, Turpeinen U, Voutilainen S, Laakso J (2016) Food and nutrient intake and nutritional status of Finnish vegans and non-vegetarians. PloS One 11(2):1–14
Escott-Stump S, Mahan LK (2000) Krause's food, nutrition and diet therapy. WB Saunders, Philadelphia, pp 1–1194
Fernstrom JD (2005) Branched-chain amino acids and brain function. J Nutr 135(6):1539S–1546S
Fernstrom JD, Fernstrom MH (2007) Tyrosine, phenylalanine, and catecholamine synthesis and function in the brain. J Nutr 137(6):1539S–1547S
Fialkow J (2016) Omega-3 fatty acid formulations in cardiovascular disease: dietary supplements are not substitutes for prescription products. Am J Cardiovasc Drug 16(4):229–239
Finckh A, Deane KD (2014) Prevention of rheumatic diseases: strategies, caveats, and future directions. Rheum Dis Clin 40(4):771–785
Fuller R editor (2012) Probiotics: the scientific basis. Springer Science & Business Media, Germany, pp 1–398
Funk JL, Frye JB, Oyarzo JN, Chen J, Zhang H, Timmermann BN (2016) Anti-inflammatory effects of the essential oils of ginger (*Zingiber officinale* Roscoe) in experimental rheumatoid arthritis. PharmaNutrition 4(3):123–131

Furrie E, Macfarlane S, Kennedy A, Cummings J, Walsh S, O'neil D, Macfarlane GT (2005) Synbiotic therapy (*Bifidobacterium longum*/Synergy 1) initiates resolution of inflammation in patients with active ulcerative colitis: a randomised controlled pilot trial. Gut 54(2):242–249
Gibson GR (2004) Fibre and effects on probiotics (the prebiotic concept). Clinical Nutrition Supplements 1(2):25–31
Gohil KJ, Patel JA (2010) A review on *Bacopa monniera*: current research and future prospects. Int J Green Pharm 4(1):1–9
Grosso G, Bei R, Mistretta A, Marventano S, Calabrese G, Masuelli L, Giganti MG, Modesti A, Galvano F, Gazzolo D (2013) Effects of vitamin C on health: a review of evidence. Front Biosci 18(3):1017–1029
Guerin M, Huntley ME, Olaizola M (2003) *Haematococcus* astaxanthin: applications for human health and nutrition. Trends Biotechnol 21(5):210–216
Gulati OP, Ottaway PB (2006) Legislation relating to nutraceuticals in the European Union with a particular focus on botanical-sourced products. Toxicology 221(1):75–87
Gupta RK, Soni P, Singh R, Shrivastava J, Rajput P (2018) Health benefits of natural nutrients: ayurveda dravya as nutraceuticals. WJPMR 4(4):253–255
Halsted CH (2003) Dietary supplements and functional foods: 2 sides of a coin? Am J Clin Nutr 77(4):1001S–1007S
Harrison FE, May JM (2009) Vitamin C function in the brain: vital role of the ascorbate transporter SVCT2. Free Radic Biol Med 46(6):719–30
Hasler CM (2002) Functional foods: benefits, concerns and challenges—a position paper from the American Council on Science and Health. J Nutr 132(12):3772–3781
Hasler CM (ed) (2005) Regulation of functional foods and nutraceuticals: a global perspective. Wiley, New Jersey, pp 1–432
Hewlings SJ, Kalman DS (2017) Curcumin: a review of its effects on human health. Foods 6(10):1–11
Hoadley JE, Rowlands JC (2014) FDA perspectives on food label claims in the United States. In: Nutraceutical and functional food regulations in the United States and around the world. Academic Press, Cambridge
Holeček M (2018) Branched-chain amino acids in health and disease: metabolism, alterations in blood plasma, and as supplements. Nutr Metab 15(1):1–12
Holick MF (2004) Sunlight and vitamin D for bone health and prevention of autoimmune diseases, cancers, and cardiovascular disease. Am J Clin Nutr 80(6):1678S–1688S
Hsu Y-W, Tsai C-F, Chang W-H, Ho Y-C, Chen W-K, Lu FJ (2008) Protective effects of *Dunaliella salina*–A carotenoids-rich alga, against carbon tetrachloride-induced hepatotoxicity in mice. Food Chem Toxicol 46(10):3311–3317
Huskisson EC (2008) Glucosamine and chondroitin for osteoarthritis. J Int Med Res 36(6):1161–1179
Irianto A, Austin B (2002) Probiotics in aquaculture. J Fish Dis 25(11):633–42
Iwata K, Inayama T, Kato T (1990) Effects of Spirulina platensis on plasma lipoprotein lipase activity in fructose-induced hyperlipidemic rats. J Nutr Sci and Vitaminol 36(2):165–171
Jain R, Pandey R, Mahant R, Rathore D (2015) A review on medicinal importance of *Emblica officinalis*. Int J Pharm Sci 6(1):72–84
Jenkins TA, Nguyen JC, Polglaze KE, Bertrand PP (2016) Influence of tryptophan and serotonin on mood and cognition with a possible role of the gut-brain axis. Nutrients 8(1):1–15
Jerosch J (2011) Effects of glucosamine and chondroitin sulfate on cartilage metabolism in OA: outlook on other nutrient partners especially omega-3 fatty acids. Int J Rheumatol 2011:1–17
Kadali SRM, Das M, Srinivasa Rao A (2014) Antidepressant activity of brahmi in albino mice. J Clin Diagn Res 8(3):35–37
Kalaiselvan S, Rasool MK (2015) The anti-inflammatory effect of triphala in arthritic-induced rats. Pharm Biol 53(1):51–60
Kalra EK (2003) Nutraceutical-definition and introduction. Aaps Pharmsci 5(3):27–28

Kalsi S, Verma SK, Neha AK, Singh NJ (2016) A review on *Glycyrrhiza glabra* (liquorice) and its pharmacological ac-tivities. Int J Pharm Sci Drug Res 4(5):234–239

Kanaley JA (2008) Growth hormone, arginine and exercise. Curr Opin Clin Nutr Metab Care 11(1):50–54

Kang S, Min H (2012) Ginseng, the 'Immunity Boost': the effects of *Panax ginseng* on immune system. J Ginseng Res 36(4):354–368

Kapoor L (2017) Handbook of Ayurvedic medicinal plants: Herbal reference library. Routledge, London, pp 1–424

Karunathilake SP, Ganegoda GU (2018) Secondary prevention of cardiovascular diseases and application of technology for early diagnosis. Biomed Res Int 2018:1–9

Katiyar SK, Raman C (2011) Green tea: a new option for the prevention or control of osteoarthritis. Arthritis Res Ther 13(4):1–2

Kavya B, Kavya N, Ramarao V, Venkateshwarlu G (2015) *Tinospora cordifolia* (Willd.) Miers: nutritional, ethno medical and therapeutic utility. Ind J Res Ayurveda Pharm 6(2):195–198

Kechagia M, Basoulis D, Konstantopoulou S, Dimitriadi D, Gyftopoulou K, Skarmoutsou N, Fakiri EM (2013) Health benefits of probiotics: a review. ISRN Nutr 2013:1–7

Kelly D, Conway S, Aminov R (2005) Commensal gut bacteria: mechanisms of immune modulation. Trends Immunol 26(6):326–333

Kemse NG, Kale AA, Joshi SR (2014) A combined supplementation of omega-3 fatty acids and micronutrients (folic acid, vitamin B12) reduces oxidative stress markers in a rat model of pregnancy induced hypertension. PLoS One 9(11):1–13

Kennedy DO (2016) B vitamins and the brain: mechanisms, dose and efficacy—a review. Nutrients 8(2):1–29

Kim JH, Yi Y-S, Kim M-Y, Cho JY (2017) Role of ginsenosides, the main active components of *Panax ginseng*, in inflammatory responses and diseases. J Ginseng Res 41(4):435–43

Kohli K, Ali J, Ansari M, Raheman Z (2005) Curcumin: a natural antiinflammatory agent. Indian J Pharmacol 37(3):141–147

Kohmoto T, Fukui F, Takaku H, Machida Y, Arai M, Mitsuoka T (1988) Effect of isomalto-oligosaccharides on human fecal flora. Bifidobacteria Microflora 7(2):61–69

Kozisek F (2005) Health risks from drinking demineralised water. Nutrients in drinking water. World Health Organization, Geneva, Switzerland

Krishnamurthy VMR, Wei G, Baird BC, Murtaugh M, Chonchol MB, Raphael KL, Greene T, Beddhu SJ (2012) High dietary fiber intake is associated with decreased inflammation and all-cause mortality in patients with chronic kidney disease. Kidney Int 81(3):300–306

Kunte KB, Kuna Y (2013) Neuroprotective effect of *Bacopa monniera* on memory deficits and ATPase system in Alzheimer's disease (AD) induced mice. J Sci Innov Res 2(4):719–735

L'Abbé MR, Dumais L, Chao E, Junkins B (2008) Health claims on foods in Canada. J Nutr 138(6):1221S–1227S

Lacy BE, De Lee R (2005) Irritable bowel syndrome: a syndrome in evolution. J Clin Gastroenterol 39(5):S230–S242

Lenoir-Wijnkoop I, Sanders ME, Cabana MD, Caglar E, Corthier G, Rayes N, Sherman PM, Timmerman HM, Vaneechoutte M, Loo Van, Wolvers DA (2007) Probiotic and prebiotic influence beyond the intestinal tract. Nutr Rev 65(11):469–489

Leray C, Wiesel ML, Freund M, Cazenave JP, Gachet C (2001) Long-chain n-3 fatty acids specifically affect rat coagulation factors dependent on vitamin K: relation to peroxidative stress. Arterioscler Thromb Vasc Biol 21(3):459–465

Lopresti AL, Drummond PD, Smith SJ (2019) A randomized, double-blind, placebo-controlled, crossover study examining the hormonal and vitality effects of ashwagandha (*Withania somnifera*) in aging, overweight males. Am J Mens Health 13(2):1–15

Lorenz RT, Cysewski GR (2000) Commercial potential for *Haematococcus* microalgae as a natural source of astaxanthin. Trends Biotechnol 18(4):160–167

Maclennan KM, Darlington CL, Smith PF (2002) The CNS effects of *Ginkgo biloba* extracts and ginkgolide B. Prog Neurobiol 67(3):235–257

Mahadevan S, Park Y (2008) Multifaceted therapeutic benefits of *Ginkgo biloba* L.: chemistry, efficacy, safety, and uses. J Food Sci 73(1):R14–R19
Mahendra P, Bisht S (2012) *Ferula asafoetida*: Traditional uses and pharmacological activity. Pharmacogn Rev 6(12):141–146
Malhotra S, Singh AP (2003) Medicinal properties of ginger (*Zingiber officinale* Rosc.). Nat Prod Rad 2(6):296–301
Manayi A, Vazirian M, Saeidnia S (2015) *Echinacea purpurea*: pharmacology, phytochemistry and analysis methods. Pharmacogn Rev 9(17):63–72
Mandal-Ghosh I, Chattopadhyay U, Baral R (2007) Neem leaf preparation enhances Th1 type immune response and anti-tumor immunity against breast tumor associated antigen. Cancer Immun 7:1–9
Manninen AH (2009) Protein hydrolysates in sports nutrition. Nutr Metab 6(1):1–5
Marteau P, Boutron-Ruault MC (2002) Nutritional advantages of probiotics and prebiotics. Br J Nutr 87(S2):S153–S157
Martini D, Del Bo C, Tassotti M, Riso P, Del Rio D, Brighenti F, Porrini M (2016) Coffee consumption and oxidative stress: a review of human intervention studies. Molecules 21(8):1–20
McKeage K, Lyseng-Williamson KA (2018) *Ginkgo biloba* extract EGb 761® in the symptomatic treatment of mild-to-moderate dementia: a profile of its use. Drugs Ther Perspect 34(8):358–366
McRae MP (2017) Dietary fiber is beneficial for the prevention of cardiovascular disease: an umbrella review of meta-analyses. J Chiropr Med 16(4):289–299
Mehmood MH, Aziz N, Ghayur MN, Gilani AH (2011) Pharmacological basis for the medicinal use of psyllium husk (Ispaghula) in constipation and diarrhea. Dig Dis Sci 56(5):1460–1471
Merly L, Smith SL (2015) Pro-inflammatory properties of shark cartilage supplement. Immunopharmacol Immunotoxicol 37(2):140–147
Mitsuoka T, Hidaka H, Eida T (1987) Effect of fructo-oligosaccharides on intestinal microflora. Food/Nahrung 31(5–6):427–436
Mukherjee PK, Rai S, Bhattacharyya S, Kumar Debnath P, Biswas TK, Jana U, Pandit S, Saha BP, Paul PK (2006) Clinical study of'Triphala'-a well known phytomedicine from India. Iran J Pharm Res 5(1):51–54
Mullaicharam A (2013) A review on evidence based practice of *Ginkgo biloba* in brain health. Int J Chem Pharm Anal 1(1):24–30
Müller O, Krawinkel M (2005) Malnutrition and health in developing countries. Can Med Assoc 173(3):279–286
Murthy KC, Vanitha A, Rajesha J, Swamy MM, Sowmya P, Ravishankar GA (2005) In vivo antioxidant activity of carotenoids from *Dunaliella salina*—a green microalga. Life Sci 76(12):1381–1390
Nagle DG, Ferreira D, Zhou YD (2006) Epigallocatechin-3-gallate (EGCG): chemical and biomedical perspectives. Phytochemistry 67(17):1849–1855
Nagpal R, Kaur A (2011) Synbiotic effect of various prebiotics on in vitro activities of probiotic lactobacilli. Ecol food Nutr 50(1):63–8
Naguib YM (2000) Antioxidant activities of astaxanthin and related carotenoids. J Agric Food Chem 48(4):1150–1154
Nasri H, Baradaran A, Shirzad H, Rafieian-Kopaei M (2014) New concepts in nutraceuticals as alternative for pharmaceuticals. Int J Prev Med 5(12):1487–1499
Neeraj VB, Johar V (2017) Bael (*Aegle marmelos*) extraordinary species of India: a review. Int J Curr Microbiol Appl Sci 6(3):1870–1887
Nemets B, Stahl Z, Belmaker R (2002) Addition of omega-3 fatty acid to maintenance medication treatment for recurrent unipolar depressive disorder. Am J Psychiatry 159(3):477–479
New SA (1999) Bone health: the role of micronutrients. Br Med Bull 55(3):619–33
Nicoletti M (2016) Microalgae nutraceuticals. Foods 5(3):1–13
Nuhu AA (2013) Spirulina (Arthrospira): an important source of nutritional and medicinal compounds. J Mar Biol 2013:1–8

Palatty PL, Haniadka R, Valder B, Arora R, Baliga MS (2013) Ginger in the prevention of nausea and vomiting: a review. Crit Rev Food Sci Nutr 53(7):659–669

Pandey M, Verma RK, Saraf SA (2010) Nutraceuticals: new era of medicine and health. Asian J Pharm Clin Res 3(1):11–15

Pandey M, Rastogi S, Rawat A (2013) Indian traditional ayurvedic system of medicine and nutritional supplementation. Evid-Based Complementary Altern Med 2013:1–12

Paniagua-Michel J (2015) Microalgal nutraceuticals. In: Handbook of marine microalgae. Elsevier, Amsterdam

Patil R, Gholave A, Jadhav J, Yadav S, Bapat V (2015) *Mucuna sanjappae* Aitawade et Yadav: a new species of Mucuna with promising yield of anti-Parkinson's drug L-DOPA. Genet Resour Crop Ev 62(1):155–162

Pereira MA, O'Reilly E, Augustsson K, Fraser GE, Goldbourt U, Heitmann BL, Hallmans G, Knekt P, Liu S, Pietinen P (2004) Dietary fiber and risk of coronary heart disease: a pooled analysis of cohort studies. Arch Intern Med 164(4):370–376

Peterson CT, Denniston K, Chopra D (2017) Therapeutic uses of Triphala in Ayurvedic medicine. J Altern Complement Med 23(8):607–614

Phillips SM, Van Loon LJ (2011) Dietary protein for athletes: from requirements to optimum adaptation. J Sports Sci 29(1):S29–S38

Pilerood SA, Prakash J (2013) Nutritional and medicinal properties of valerian (*Valeriana officinalis)* herb: a review. IJFND 1(1):1–9

Pingale Shirish S (2010) Hepatoprotection study of leaves powder of *Azadirachta indica. A. Juss.* Int J Pharm Sci Rev Res 3(2):37–42

Pozzilli P, Guglielmi C, Caprio S, Buzzetti R (2011) Obesity, autoimmunity, and double diabetes in youth. Diabetes Care 34(2):S166–S170

Prakash P, Gupta N (2005) Therapeutic uses of *Ocimum sanctum* Linn (Tulsi) with a note on eugenol and its pharmacological actions: a short review. Indian J Physiol Pharmacol 49(2):125–131

Prasad S, Sung B, Aggarwal BB (2012) Age-associated chronic diseases require age-old medicine: role of chronic inflammation. Prev Med 54:S29–S37

Pulz O, Gross W (2004) Valuable products from biotechnology of microalgae. Appl Microbiol Biotechnol 65(6):635–648

Rahman S, Ansari RA, Rehman H, Parvez S, Raisuddin S (2011) Nordihydroguaiaretic acid from creosote bush (*Larrea tridentata*) mitigates 12-O-tetradecanoylphorbol-13-acetate-induced inflammatory and oxidative stress responses of tumor promotion cascade in mouse skin. Evid Based Complement Alternat Med 2011:1–10

Ramamurthy V, Raveendran S, Kumar HA (2014) Effect of *Lyngbya Majuscula* on the immunity and survival of *Aeromonas Hydrophilia* infected *Mugil Cephalus*. Int J Innovres Sci Eng Technol 3(2):9782–9789

Ramchander J, Middha A (2017) Recent advances on Senna as a laxative: A comprehensive review. J Pharmacogn Phytochem 6(2):349–53

Rao RV, Descamps O, John V, Bredesen DE (2012) Ayurvedic medicinal plants for Alzheimer's disease: a review. Alzheimers Res Ther 4(3):1–9

Reid I, Bristow S, Bolland MJ (2015) Calcium supplements: benefits and risks. J Intern Med 278(4):354–382

Russo A, Borrelli F (2005) *Bacopa monniera*, a reputed nootropic plant: an overview. Phytomedicine 12(4):305–317

Saha S, Ghosh S (2012) *Tinospora cordifolia*: one plant, many roles. Anc Sci Life 31(4):151–159

Saha S, Murray P (2018) Exploitation of microalgae species for nutraceutical purposes: cultivation aspects. Fermentation 4(2):1–17

Saito M (2007) Role of FOSHU (food for specified health uses) for healthier life. Yakugaku Zasshi 127(3):407–16

Santini A, Cammarata SM, Capone G, Ianaro A, Tenore GC, Pani L, Novellino E (2018) Nutraceuticals: Opening the debate for a regulatory framework. Br J Clin Pharmacol 84(4):659–72

Schaafsma G (2000) The protein digestibility–corrected amino acid score. J Nutr 130(7):1865S–1867S

Schultz M, Munro K, Tannock G, Melchner I, Göttl C, Schwietz H, Schölmerich J, Rath H (2004) Effects of feeding a probiotic preparation (SIM) containing inulin on the severity of colitis and on the composition of the intestinal microflora in HLA-B27 transgenic rats. Clin Diagn Lab Immunol 11(3):581–587

Schwedhelm E, Maas R, Freese R, Jung D, Lukacs Z, Jambrecina A, Spickler W, Schulze F, Böger RH (2008) Pharmacokinetic and pharmacodynamic properties of oral L-citrulline and L-arginine: impact on nitric oxide metabolism. Br J Clin Pharmacol 65(1):51–59

Sharma Y (2017) Ginger (*Zingiber officinale*)-an elixir of life a review. Pharma Innovation 6(10):22–27

Sharma V, Sharma S, Pracheta Paliwal R (2011) *Withania somnifera*: a rejuvenating ayurvedic medicinal herb for the treatment. Int J Pharmtech Res 3(1):187–192

Shen CL, Smith BJ, Lo DF, Chyu MC, Dunn DM, Chen CH, Kwun IS (2012) Dietary polyphenols and mechanisms of osteoarthritis. J Nutr Biochem 23(11):1367–1377

Shieh D-E, Liu L-T, Lin CC (2000) Antioxidant and free radical scavenging effects of baicalein, baicalin and wogonin. Anticancer Res 20(5A):2861–2865

Tajiri K, Shimizu Y (2013) Branched-chain amino acids in liver diseases. World J Gastroenterol 19(43):7620–7629

Shimomura Y, Yamamoto Y, Bajotto G, Sato J, Murakami T, Shimomura N, Kobayashi H, Mawatari KJ (2006) Nutraceutical effects of branched-chain amino acids on skeletal muscle. J Nutr 136 (2):529S–532S

Shin Y-W, Bae E-A, Lee B, Lee SH, Kim JA, Kim Y-S, Kim D-H (2007) In vitro and in vivo antiallergic effects of *Glycyrrhiza glabra* and its components. Planta Med 73(03):257–261

Shinde N, Bangar B, Deshmukh S, Kumbhar P (2014) Nutraceuticals: a review on current status. Res J Pharm Tech 7(1):110–113

Shoba G, Joy D, Joseph T, Majeed M, Rajendran R, Srinivas PS (1998) Influence of piperine on the pharmacokinetics of curcumin in animals and human volunteers. Planta Med 64(04):353–356

Singh R, Geetanjali (2013) Nutraceuticals: promising health product. Int Res J Med Sci 1(1):14–17

Sotodeh-Asl N, Tamadon M-R, Malek F, Zahmatkesh M (2014) Vitamin D deficiency and psychological disorders. J Parathyr Dis 2(1):21–25

Stanislavov R, Rohdewald P (2014) PACR (pine bark extract, L-arginine, L-citrulline, rose hip extract) improves emotional, physical health and sexual function in peri-menopausal women. J Women's Health Care 3(6):1–6

Steed H, Macfarlane S (2009) Mechanisms of prebiotic impact on health. Prebiotics and Probiotics Science and Technology, Springer, New York

Stoll AL, Severus WE, Freeman MP, Rueter S, Zboyan HA, Diamond E, Cress KK, Marangell LB (1999) Omega 3 fatty acids in bipolar disorder: a preliminary double-blind, placebo-controlled trial. Arch Gen Psychiatry 56(5):407–412

Suganya T, Varman M, Masjuki H, Renganathan S (2016) Macroalgae and microalgae as a potential source for commercial applications along with biofuels production: a biorefinery approach. Renew Sust Energ Rev 55:909–41

Suvarna V, Boby VU (2005) Probiotics in human health: A current assessment. Curr Sci 88(11):1744–1748

Swinburn BA, Caterson I, Seidell JC, James W (2004) Diet, nutrition and the prevention of excess weight gain and obesity. Public Health Nutr 7(1a):123–146

Sýkora J, Valecková K, Amlerová J, Siala K, Dedek P, Watkins S, Varvarovská J, Stožický F, Pazdiora P, Schwarz J (2005) Effects of a specially designed fermented milk product containing probiotic *Lactobacillus casei* DN-114 001 and the eradication of *H. pylori* in children: a prospective randomized double-blind study. J Clin Gastroenterol 39(8):692–698

Tajiri K, Shimizu Y (2013) Branched-chain amino acids in liver diseases. World J Gastroenterol 19(43):7620–7629

Tapsell LC (2008) Evidence for health claims: A perspective from the Australia-New Zealand region. J Nutr 138(6):1206S–1209S
Tee ES, Tamin S, Ilyas R, Ramos A, Tan WL, Lai DKS, Kongchuntuk H (2002) Current status of nutrition labelling and claims in the South-East Asian region: are we in harmony? Asia Pac J Clin Nutr 11(2):S80–S86
Turner T, Burri BJ (2013) Potential nutritional benefits of current citrus consumption. Agriculture 3(1):170–187
Ünlü C, Daniels L, Vrouenraets BC, Boermeester MA (2012) A systematic review of high-fibre dietary therapy in diverticular disease. Int J Colorectal Dis 27(4):419–427
Ushiyama A, Tanaka K, Aiba Y, Shiba T, Takagi A, Mine T, Koga Y (2003) Lactobacillus gasseri OLL2716 as a probiotic in clarithromycin-resistant Helicobacter pylori infection. J Gastroen Hepatol 18(8):986–991
Vasiliadis HS, Tsikopoulos K (2017) Glucosamine and chondroitin for the treatment of osteoarthritis. World J Orthop 8(1):1–11
Veldurthy V, Wei R, Oz L, Dhawan P, Jeon YH, Christakos S (2016) Vitamin D, calcium homeostasis and aging. Bone Res 4:1–7
Verma A, Mogra R (2013) Psyllium (*Plantago ovata*) husk: a wonder food for good health. Int J Sci Res 4(9):1581–1585
Verma B, Popli H (2018) Regulations of nutraceuticals in India & us. Pharma Innovation 7(7):811–816
Wang J, Zhang X, Lan H, Wang W (2017) Effect of garlic supplement in the management of type 2 diabetes mellitus (T2DM): a meta-analysis of randomized controlled trials. Food Nutr Res 61(1):1–10
Wärnberg J, Marcos A, Bueno G, Moreno LA (2009) Functional benefits of psyllium fiber supplementation. Curr Top Nutraceutr 7(2):1–10
Wąsik A, Antkiewicz-Michaluk LJPR (2017) The mechanism of neuroprotective action of natural compounds. Pharmacol Rep 69(5):851–860
Whitman MM (2001) Understanding the perceived need for complementary and alternative nutraceuticals: lifestyle issues. Clin J Oncol Nurs 5(5):190–194
Wu S, Silverberg J, Joks R, Durkin H, Smith-Norowitz T (2012) Green tea (*Camelia sinensis*) mediated suppression of IgE production by peripheral blood mononuclear cells of allergic asthmatic humans. Scand J Immunol 76(3):306–310
www.assocham.org/newsdetail.php?id=6259
www.bccresearch.com/market-research/food...nutraceuticals-global-markets
www.figlobal.com
www.grandviewresearch.com/industry-analysis/nutraceuticals-market
www.mordorintelligence.com
Xing LC, Santhi D, Shar AG, Saeed M, Arain MA, Shar AH, Bhutto ZA, Katar M, Manzoor R, El Hack M (2017) Psyllium husk (*Plantago ovata*) as a potent hypocholesterolemic agent in animal, human and poultry. Int J Pharmacol 13:690–697
Yang Y (2008) Scientific substantiation of functional food health claims in China. J Nutr 138(6):1199S–205S
Yuan JP, Peng J, Yin K, Wang JH (2011) Potential health-promoting effects of astaxanthin: A high-value carotenoid mostly from microalgae. Mol Nutr Food Res 55(1):150–65
Zeisel SH (1999) Regulation of “nutraceuticals”. Science 285(5435):1853–1855
Zink A, Traidl-Hoffmann C (2015) Green tea in dermatology–myths and facts. J Dtsch Dermatol Ges 13(8):768–775

Nitika Kapoor Masters in Human Genetics. Awardee of ICMR Junior and Senior research fellowships. Presently pursuing Ph.D. at CSIR-Indian Institute of Integrative Medicine, Jammu. Working on regulatory aspects of secondary metabolism in medicinal plants, employing cutting edge molecular tools, transcriptomics, metabolomics, etc.

Vijay Lakshmi Jamwal Masters in Human Genetics. Presently pursuing Ph.D. at AcSIR. Working at CSIR-Indian Institute of Integrative Medicine (Jammu), towards understanding the physiological and ecological roles of secondary metabolites in high value medicinal plants. Trained in molecular tools such as CRISPR-Cas9, transcriptomics, metagenomics, metabolomics, transgenics, etc.

Manish R. Shukla With over seventeen years of experience in contributing and leading various research and commercial areas of Biotechnology worked as a Senior Scientist in TIFR, Mumbai establishing various protein analytics platforms and researched on various aspects of Algal stress metabolism. During his PhD from MSU Vadodara and Postdoctoral studies at Louisiana State University, USA, under the aegis of the NASA NSF, added deeper understanding in the areas of bacterial radiation resistance and their potential to Mars survival. In Reliance R&D, he has established platforms to develop robust understandings on the metabolomics and proteomics of Algae, to harness and optimize their productivity for commercial use. Further Dr Shukla is driving the potential of Algal technology proficiency at Reliance R&D for various food and Ingredients segment.

Sumit G. Gandhi Working as senior faculty (Principal Scientist) at CSIR-IIIM (Jammu), on various aspects of secondary metabolism in microorganisms and medicinal & aromatic plants, since more than a decade. PhD (JNU, Delhi) at CCMB (Hyderabad) focused on understanding molecular effectors and mechanisms of epigenetic regulation. Bachelors & Masters in Microbiology. Dr. Gandhi has resourcefully collaborated with industries and academia, on various projects and has been a recipient of grants from public and private sectors.

Biofuels

Algae Biodiesel: Fundamentals and Future Prospects

Ranjana Bhati

Abstract Global economy has been benefited from fossil fuels in the past but at the cost of increasing CO_2 level and other environmental hazards. Fast-growing concern about global warming, exhaustion of nonrenewable energy reserves and rising cost of petroleum-derived fuels, led to quest for other sustainable renewable energy alternatives. In current scenario, renewable and environment-friendly biofuel production is the only sustainable alternative to shrinking petrodiesel reserves. Over the years, biodiesel has grabbed the most focus as a potential liquid biofuel. First-generation biodiesel was produced mainly from edible vegetable oils and provoke large number of debate, primarily owing to competition with overall food production. Subsequently, second-generation biodiesel was produced by using nonedible oil sources like karanja, jatropha and mahua oils. First- and second-generation biodiesel have demerits, mainly the expensive set up, land requirements for plants cultivation and competition with net food production. Moreover, biodiesel derived from these sources could not practically fulfill the small part of current transport fuels requirements. Thus, the focus of researchers has been shifted to the third-generation biofuel from microalgae, which is highly promising. Utilization of microalgae for different types of renewable biofuel production is having diverse benefits of overcoming the energy crisis and environmental pollution control. Microalgae are photosynthetic eukaryotes and capable to grow in different growth conditions with CO_2 biofixation. Microalgae are 10–50 times efficient in solar energy capture than plants and also higher in biomass accumulation than energy crops. Algae can grow in diverse aquatic habitat and on land that is agriculturally barren, therefore, no arable land competition for food and feed. The key attraction of biodiesel production with microalgae lies in their ability to tolerate high CO_2 concentration. Moreover, wastewater (rich in nutrients) can be effectively utilized for microalgae cultivations and subsequently results in wastewater recycling. Biodiesel from algae appears to be the most economic and sustainable energy source with the CO_2 utilization and wastewater recycling.

Keywords Algae · Biofuel · Biodiesel · Lipid productivity · Photobioreactors

---

R. Bhati (✉)
Department of Microbiology, Bundelkhand University, Jhansi, Uttar Pradesh 284128, India

A. Saxena (ed.), *Biotechnology Business - Concept to Delivery*, EcoProduction,
https://doi.org/10.1007/978-3-030-36130-3_5

1 Introduction

Ever-increasing energy demand, concern about climate change and exhaustion of fossil fuels has driven the interest to look for eco-friendly and sustainable energy resources. Due to environmental hazards imposed by the elevated CO_2 level in the atmosphere, United Nations (1997) introduced the Kyoto Protocol to lower the collective greenhouse gases emission by 5.2% as compared to the emission in 1990, and this protocol was ratified by more than 170 countries (Gutiérrez et al. 2008; Wang et al. 2008; Ho et al. 2011). In addition, 'carbon credit' system with an approximate unit price of US $270/ton was proposed by United Nations in 2010 (Stewart and Hessami 2005). Further at the climate change convention (Copenhagen 2010), nations decided to grant about US $100 billion for greenhouse reduction by 2020 (Kintisch 2010; Ho et al. 2011).

Over the last few years, immense uses of fossil fuels to fulfill the energy demand impose serious environmental hazards by rising CO_2 concentration in the atmosphere. In 2012, total CO_2 emission was 36% of total global emissions generated by the utilization of liquid fuels and predicted that by the year 2035, emission of CO_2 would be doubled (EIA 2016). Owning to this, the European Union Renewable Energy Directive (RED) advocates that by 2020, 15% of the total energy demand supplied to the UK should be derived from renewable resources (Gul 2016; Adeniyi et al. 2018). Currently, 80% of energy requirement is derived from fossils fuels. According to these energy requirements, it is projected that after 2050, conventional oil reserves will be completely exhausted (Demirbas and Demirbas 2011; Ullah et al. 2014). So, the development of carbon neural, renewable and eco-friendly energy resources to overcome the energy crisis is the need of the hour. Research and development efforts are now, therefore, directed to store, fix and utilize CO_2 in order to minimize its release into the atmosphere. The potential liquid biofuel that has attracted the most attention recently is biodiesel (Chisti 2007).

This chapter gives a detailed insight on algae as a potent host for biodiesel production. The chapter focuses on different microalgae biodiesel production aspects that lead to sustainable and cost-effective production process. Various cultivation systems of algae, harvesting and downstream processes for biomass recovery are discussed in detail. Lipid content with biodiesel production and major biofuel manufacturers' are described. Detailed insight on technical challenges and economic feasibility of commercial algal biodiesel production process are discussed with cost comparison to fossil fuels.

2 Biodiesel: Introduction

Biodiesel is chemically monoalkylester made by transesterification reaction of free fatty acids with alcohols and obtained from renewable resources like vegetable oils and animal fats. Biodiesel is ecologically less toxic than petroleum-based diesel and

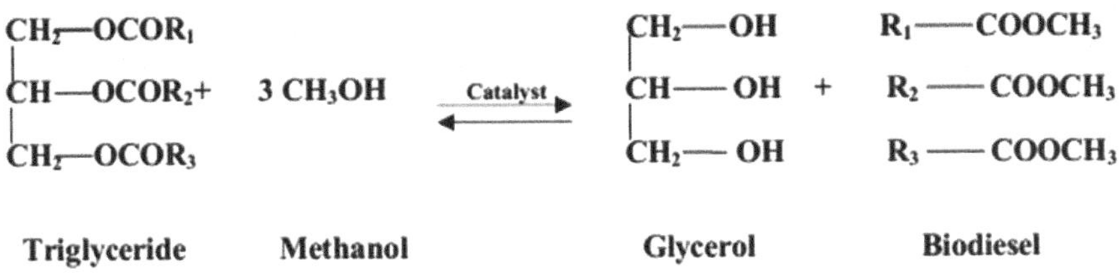

Fig. 1 Algal biodiesel production by transesterification, $R_{1\text{-}3}$ are hydrocarbon group (Chisti 2007)

has chemical and physical properties similar to petrodiesel (Chisti 2007; Demirbas 2011; Adeniyi et al. 2018) (Fig. 1).

Biodiesel has received immense interest in past few years as the sustainable choice to petroleum diesel fuel due to its renewability, nontoxic and eco-friendly characteristics (Krawczyk 1996). Biodiesel has comparable conventional diesel engine performance to petrodiesel and reduces the emissions of pollutants (100% less sulfur dioxide, 84% less particulate matter, 46% less carbon monoxide and 37% less unburned hydrocarbons emissions) as compared to conventional diesel fuel (McMillen et al. 2005).

The first-generation biodiesel is produced from feedstock of edible oils like rapeseed and soybean, generated a lot of conflicts due to food and feed competitions. The second-generation biodiesel is produced by using nonedible oil sources. The first- and second-generation biodiesel based on terrestrial plants are potentially competing with net food production (Chisti 2008). The focus of researchers has now been shifted to the next-generation biodiesel. The third-generation biodiesel is both promising and different, and it is based on simple microscopic organisms that live in aquatic habitat, i.e., microalgae.

3 Microalgae as Host for Biodiesel Production

Microalgae are receiving much importance as a potential candidate for CO_2 sequestration and alternative host for renewable energy production. Algae are photosynthetic organisms that can be cultivated in various environmental conditions with ability to fix CO_2. The benefits of using microalgae for biofuels production lie in their capacity to tolerate high CO_2 concentration, allowing efficient CO_2 capturing (CO_2 content 5–15%) (Hsueh et al. 2007). On an estimate, 1.83 kg of CO_2 is utilized for the production of 1 kg of algal biomass (dry cell weight) (Chisti 2007). Algae have high solar energy capture efficiency, faster growth and higher biomass production in comparison with energy crops (Wang et al. 2008; Brennan and Owende 2010). Algae can grow in diverse aquatic habitat and on land that is agriculturally barren. Therefore, algae do not affect the crop cultivation system and no arable land competition for feed and food (Brown and Zeiler 1993; Aresta et al. 2005). Moreover, algae can successfully utilize and grow in high nutrient concentrations of wastewater and have dual potential for waste treatment as they can utilize waste nutrients for growth and in turn eliminate N and P from the wastewaters (Mallick 2002). This indicates that algae are noncompetitive with other users for freshwater, thus reducing the nutrients costs and preserving the valuable freshwater resources (Campbell 2008) (Fig. 2).

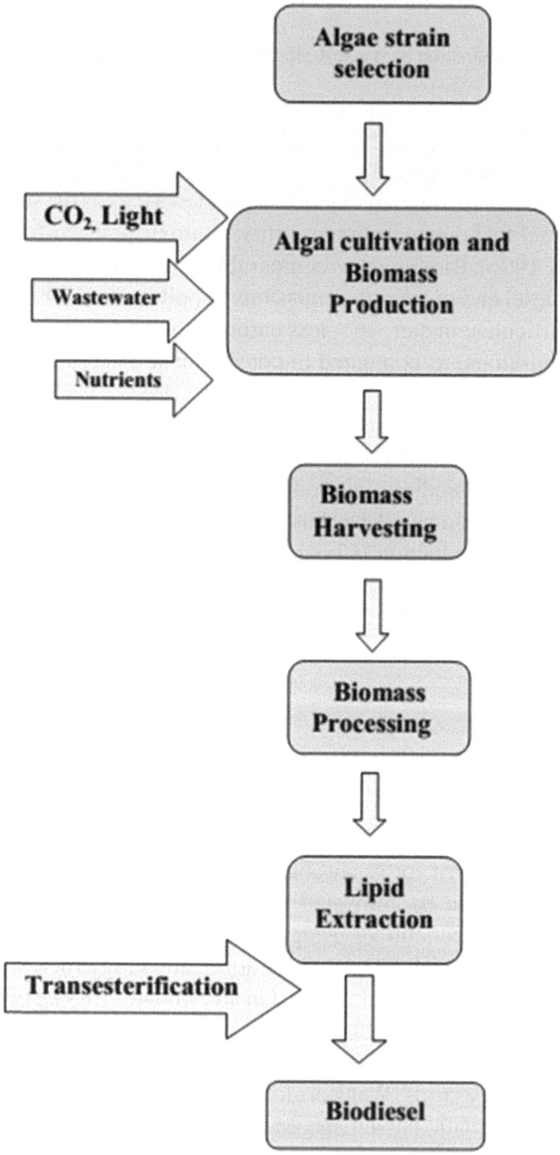

Fig. 2 Overview of biodiesel production with algae

4 Microalgae Lipid Content

Different microalgae species accumulate distinct ratios of proteins, lipids and carbohydrates. Microalgae lipid accumulation ought to be high to attain the economic performance (Xu et al. 2006). Most algal species can be triggered to synthesize sizeable amount of lipids, resulting in high oil yield. The typical reported lipid content ranges from 1 and 70%; however, with specific growth conditions, lipid accumulation reached up to 90% (dcw) (Chisti 2007; Li et al. 2011). Lipid productivity in many microalgal species can be increased by physiological stresses such as nutrient deficiencies/limitations, varying physical parameters (light intensity, pH temperature), mode of nutrition (phototrophic, heterotrophic and photoheterotrophic mixotrophic) and optimization of medium composition (carbon substrate, salts, phosphorus, nitrogen and vitamins). Lipids are then converted to biofuel through a transesterification reaction (Chisti 2007). Brown et al. (1969) reported an oil yield of 86% (dcw) from a bloom of *Botryococcusbraunii* in a freshwater lake. But a very slow growth rate is the main hindrance in exploring *B. braunii* as the commercial organism for the production of biodiesel (Dayananda et al. 2007). *Scenedesmus*sp. strain JPCC GA0024 was found to accumulate lipid up to 73% (dcw) lipid content when grown in artificial seawater (Matsunaga et al. 2009). Glucose supplementation in growth media resulted in lipid accumulation up to 61% (dcw) by *Chlamydomonas reinhardtii* CC1010 (Karpagam et al. 2015). Extensive research has been done for enhancing the lipid production by varying the cultural and physical conditions. Table 1 summarizes some significant attempts to increase the lipid yield of microalgal species with different growth conditions.

5 Microalgae Cultivation

Algae can grow anywhere with abundant sunlight, carbon source, water, micro and macronutrients with optimum temperature. Algal cultivation for biomass has several advantages like higher oil percentage than other sources. Second, production of extremely large amount of biomass and this algal biomass production has limited food and feed competition in market. Moreover, algae can grow with brackish water, seawater and freshwater resources (Campbell 2008).

For enhanced production of biodiesel, type of cultivation system for the growth of algae is a significant factor. Cultivation systems for algal growth are broadly divided into types: open-pond system and closed controlled systems. Open-pond system is the oldest, simplest and cost-effective method for mass cultivation of algae. Raceway pond is defined as a closed oval loop, ranges in depth from 20 to 30 cm. They are incorporated with paddle wheel to mix the water and avoid culture settling and minimizing the shading effect. These are generally kept shallow for sufficient sunlight penetration. There are several disadvantages associated with raceway pond: little productivity as compared to photo bioreactors (Brennan and Owende 2010; Rawat et al.

Table 1 Increase lipid content of diverse microalgae species with different growth conditions

Microalgal species	Lipid content (% dry cell weight)	Growth conditions	References
Botryococcusbraunii	86	Nitrogen limitation	Brown et al. (1969)
Chlorella vulgaris	53	Nitrogen limitation	Piorreck and Pohl (1984)
Chlorella protothecoides	58	Heterotrophy (0.1% glucose with reduced nitrogen)	Miao and Wu (2004)
Chlorella sp.	38	Heterotrophy (0.1% glucose with sodium thiosulphate supplementations)	Feng et al. (2005)
Chlorellaprotothecoides	55	Heterotrophy with corn powder hydrolysate + nitrogen limitation	Xu et al. (2006)
*Scenedesmus*sp. strain JPCC GA0024	73	Phototrophy with artificial seawater	Matsunaga et al. (2009)
Scenedesmusobliquus	39	Carbon dioxide	Ho et al. (2010)
Chlorella zofingiensis	55	Nitrogen limitation	Feng et al. (2011)
Chlorella vulgaris ESP-31	56	Photoautotrophic	Yeh and Chang (2011)
Chlorella vulgaris	57	Nitrogen, phosphorus and iron limitations	Mallick et al. (2012)
Chlorella sorokiniana	32	Glucose	Li et al. (2013)
Chlorella sp. HQ	63	Photoautotrophic	Zhang et al. (2014)
Chlorococcumlittorale	48	Carbon dioxide	Ota et al. (2015)
*Chlamydomonasreinhardtii*CC1010	61	Glucose	Karpagam et al. (2015)
Scenedemusobliquus (SAG 276-10)	34	Carbon dioxide with wastewater	Arbib et al. (2017)
Dunaliellatertiolecta	21	Carbon dioxide with Nacl and NaOH supplementation	Kumar et al. (2018)
Selenastrum sp. GA66	49	Nacl and nitrate supplementation	Chakravarty and Mallick (2019)

2013; Slade and Bauen 2013). Other demerits are poor light utilization, uncontrolled temperature, contamination by predator and heterotrophs, evaporation loss and CO_2 dispersion into the atmosphere. However, cultivation with open-pond system aids in wastewater treatment and requires fewer amounts of energy and manpower (Ugwu et al. 2008; Brennan and Owende 2010).

Closed controlled systems are mainly photobioreactors of diverse shapes as flat plate, tubular and column. In PBRs, the culture medium and required nutrients for algal growth are circulated from a central reservoir and light, and pH level is maintained in a controlled process. Though the PBR offers improved control of culture conditions, they are much costlier than natural cultivation. Auxiliary energy demand and capital cost are also higher (Chisti 2007; Brennan and Owende 2010; Slade and Bauen 2013).

To overcome the disadvantages associated with these two cultivation systems, hybrid system is introduced. The hybrid system is a two-stage cultivation method coupling the different growth stages of photobioreactor with open ponds. Higher biomass concentration and culture purity maintenances are achieved with photobioreactor during first growth stage. The second growth stage is attempted with raceway pond in which algal cells are subjected to nutrient limitations, which boost up the desired lipid accumulation (Huntley and Redalje 2007; Rodolfi et al. 2008). Raceway ponds are suitable as algal cells are subjected for natural environmental stress to increase lipid yield.

6 Harvesting Methods of Microalgae Biomass

Removal of algae biomass from broth is known as biomass harvesting. It is a difficult and problematic task due to microscopic size of algal cells (Grima et al. 2003). A suitable harvesting method is required to remove large water content and processing of high volumes of algal biomass since microalgae broth is relatively diluted (0.5–5 kg m^{-3} dry weight). Choice of best harvesting method depends on microalgal characteristics, e.g., size, density of microalgae and significance of the desired products (Grima et al. 2003; Rawat et al. 2013). 20–30% the total biomass production cost is incurred solely by algal biomass recovery from broth (Mata et al. 2010). Biomass processing for biodiesel production is a significant step which influences the overall usefulness of product. For economical biodiesel production, suitable, energy-efficient and cost-effective harvesting methods are required. Centrifugation, filtration, ultrafiltration, sedimentation and flocculation are the most common harvesting methods. Belt filtering, microstraining, sedimentation and flotation with float collection are the four harvesting methods studied by Weissman and Goebel (1987). These methods principally work on difference on a size and density for biomass separation.

Flocculations increase the particle size and accelerate the sedimentation process. Sedimentation process can be fastened by the addition of chemical flocculants such as ferric chloride, alum and lime, but they are costly for large-scale operations and not eco-friendly sustainable alternatives. Nontoxic and low-cost flocculants could be the best choices. Due to these demerits, flocculation is not widely used method for cost-effective and efficient production (Grima et al. 2003; Greenwell et al. 2010; Rawat et al. 2013).

Centrifugation is the most frequently used method for harvesting of algal biomass where quick algal biomass separation is performed by increase in gravitational acceleration, thereby resulting in harvesting biomass with >95% efficiency at 13000 g (Greenwell et al. 2010). As a result, for 15% total suspended solids, 150 times increase in slurry concentration is practically feasible (Mohn 1980). This technique is easy, quick, energy-intense, efficient and nondisruptive for high-value intracellular products. The disadvantages associated with the process are high power expenses and maintenance requirements, consequently increasing the production cost.

Conventional filtration is the most appropriate harvesting process for big size algae such as *Spirulina*. Principally, it works under pressure and filtration aided by cellulose or diatomaceous earth to enhance the effectiveness (Mohn 1980). Membrane microfiltration and ultrafiltration (a type of membrane filtration with hydrostatic pressure) are technically feasible alternatives to conventional filtration and used for smaller size algae suspension (<30 mm) (Petrusevski et al. 1995; Brennan and Owende 2010). Microfiltration is more competent and appropriate method for fragile microalgal cells harvesting (Grima et al. 2003). The key limitations of membrane filtration processes are membrane replacement cost and pumping in larger production scales results in high cost of process. Membrane filtration' operational costs are low as compared to centrifugation technique, thus making the technology more lucrative (Greenwell et al. 2010).

Gravity sedimentation is the most frequently used and simplest harvesting technique for algae biomass treatment in large volumes of water and wastewater cultures. However, process is accepted as suitable method for large microalgae such as *Spirulina.* The process is cost effective, simple, and only a settling tank is required for large-scale biomass harvesting. The major disadvantages associated with this process are time consumption, ineffective for small size algae and separation required relatively longer settling time (Grima et al. 2003; Chen et al. 2011). Due to high energy requirement and operation cost, filtration and centrifugation appear to be very far from commercial applications. However, mass harvesting methods like flocculation offer a promising alternative with lower energy input and low cost.

7 Biodiesel Production

For biodiesel production, microalgal biomass needs to be processed for lipids and fatty acids extraction. Several methods are used by researchers for extraction of lipids. Extraction of lipid is expensive and one of the extensively discussed processes for biodiesel production. Large number of microalgae grown in diverse aquatic habitat produced neutral lipids due to their lower degree of unsaturation. These lipids are ideal candidate for conversion to biodiesel (Viswanath et al. 2010).

Choice of the suitable lipid extraction methods depends on accuracy, efficiency, cost effectiveness, ease to carry out, reproducibility and precision of the method employed. First algal cells must be disrupted chemically or mechanically for lipid extraction. Bligh and Dyer method, Folch method and gravimetric method are widely reported for lipid extraction. A number of effective chemical solvent extraction methods such as ethanol, methanol, *n*-hexane and mixed polar/nonpolar chemical solvents (e.g., chloroform/methanol) are also available, and efficiency of extraction is reliant on algal species (Halim et al. 2012).

A novel method for lipid extraction from diluted algal cultures employing switchable polarity solvents (SPS) was developed by Samorì et al. (2013). Addition of CO_2 into the extracting SPS results in 70–80% efficient lipid recovery. Use of wet algal biomass for lipid extraction reduces the power consumption required for dewatering process of biomass. Ideal lipid extraction method should be highly specific and selective for algal lipids and should circumvent the co-extraction of other different compounds like carbohydrates and protein (Halim et al. 2012). The Bligh and Dyer's (1959) lipid extraction method is the best and extensively used method for total lipid extraction from microalgae where precipitation of proteins takes place in the middle of the two distinct liquid phases. This is widely accepted method and used for large-scale lipid extraction (Rawat et al. 2013).

After lipid extraction processes, algal oil is converted into biodiesel through transesterification. Biodiesel is mono-esters that are produced by transesterification, a chemical reaction between triglycerides and alcohol in the presence of a catalyst (Chisti 2007; Demirbas 2011; Adeniyi et al. 2018). Transesterification is commonly used to reduce the viscosity of bio-oils and converting them to biodiesel. Since crude algal oil is highly viscous, so their conversion to fatty acid alkyl esters (low in molecular weight) is required (Demirbas 2008). Due to gravity separation, two layers are formed at the end of the end of the reaction, biodiesel is separated at upper layer (the main product), while glycerol is at bottom layer (by-product). This process was followed by several purification steps like washing with water and evaporation to obtain high purity and quality biodiesel (Demirbas 2008).

Culture conditions play a significant role in determining the lipid and fatty acid contents of microalgae. Algae contain both saturated and monounsaturated fatty acids, and high proportion of these fatty acids is desirable for good quality of fuel (Sheehan et al. 1998; Demirbas 2011). Finally, a number of companies are investing in algae biofuel research, funding on the research and development in order to make biofuel from algae—a commercial reality. Table 2 presents a list of algae biofuel manufacturers with location.

8 Techno-Economic Analysis

Economic concern and achieving the desired revenue are the important factors for any production plant. Algae biodiesel presents a promising environment-friendly production system; however, a number of barriers should be taken into account. These

Table 2 Major algae biofuel producing companies (Chisti and Yan 2011)

Manufacturers	Biofuel types	Locations	Website
AFS Biooil^{TM}	Biodiesel	South San Francisco, USA	www.afsbiooil.com
Algenol Biofuels	Biodiesel, ethanol, gasoline, jet fuel	Florida, USA	www.algenol.com
Bionavitas, Inc.	Biofuel	Redmond, Washington USA	www.bionavitas.com
Petroalgae Inc.	Algae oil	New York, USA	www.petroalgae.com
Poet LLC	Bioethanol	South Dakota, USA	www.poet.com
Sapphire Energy	Biofuel, green crude	San Diego, Californian USA	sapphireenergy.com
Solix Biofuels	Biofuel	Fort Collins, USA	www.solixbiofuels.com
Seambiotic Ltd.	Biodiesel, ethanol	Israel	www.seambiotic.com
Synthetic genomics incorporation	Biodiesel	La Jolla, CA, USA	www.syntheticgenomics.com
Terra Via Holdings, Inc (formlySolazyme)	Algae oil	South San Francisco, USA	http://terravia.com

barriers include high energy demand (to supply water and CO_2 and for mixing), heat requirement (for biomass drying) and nutrient requirements (carbon, phosphorus and nitrogen) (Beal et al. 2015). Biodiesel production process needs to be efficient by reducing the huge cost incurred by biomass production and downstream processing. Further, biodiesel cost depends on choice of algal strain, oil content, microalgae yield, over head cost, plant capacity, plant location and design. Biodiesel production with soya oil or rapeseed enhances the cost of production since feedstock cost is the largest expense. Further, considerable reduction in biodiesel cost is obtained by exploitation of feedstock that is nonedible like jatropha oil. By financial analysis of biodiesel production cost, it has been found that ethyl esters production costs from jatropha is to be around 0.40 € per litre (Nevase et al. 2012), however ethyl esters production costs from palm oil is 0.57 € per litre (Chisti 2007)

It has been found that the biomass production costs is the only relevant factor after comparative evaluation of microalgae biodiesel production from photobioreactors and raceways. Capital investments for open-pond systems are ten times less than photobioreactors (Chisti 2007). Biomass recovery cost for broth produced in raceways is much higher than the cost of biomass recovery for broth produced by photobioreactors. The concentration of biomass produced in raceways is approx 30 times less than concentration of biomass produced in photobioreactors (Chisti 2007). In open ponds, typical microalgae productivity is 30–50 t/ha/y (Benemann and Oswald 1996; Sheehan et al. 1998). For open-pond systems and photobioreactors, estimated cost of production from algal biomass is ($10/kg) and ($30–$70/kg), respectively.

When compared with conventional agricultural biomass, this cost is higher in two and more than three order of magnitude, respectively (Carlsson et al. 2007). Presume that 30% of biomass weight is oil, free of cost CO_2 is available (flue gas), and estimated production cost for one liter of oil from raceway ponds and photobioreactors is \$1.81 and \$1.40, respectively (Chisti 2007). Moreover, for biodiesel to be cost competitive with petrodiesel, algal oil cost should be less than \$0.48/L (Chisti 2007; Demirbas and Demirbas 2011). Despite the high biomass productivity, algae biodiesel production is still not economically competitive with petrodiesel, and these costs are too high to address the current energy market.

9 Conclusion and Future Perspectives

To fulfill the growing energy requirement of mankind, algae biodiesel is cost effective, biodegradable and renewable alternative while sustaining growth. Algae are the fastest-growing eukaryotes containing 50% of oil by weight. It is noteworthy here that a number of reports for biodiesel production at laboratory scale or pilot scale are available, but reports for large-scale studies are scarce. Though closed systems are quite successful for biodiesel production, large-scale production and operation of closed systems are the need of the hour.

A multidisciplinary approach involving mass algal cultivation with the utilization of inexpensive sources like wastewaters, and CO_2 from flue gas, coupled with the extraction of value-added products is required make the process cost effective and more economical. Most importantly, lucrative and power-competent harvesting methods are also necessary to make the entire biodiesel production process more efficient. Furthermore, algal biofuel production systems are in an early development stage, and for approaching commercial profitability, extensive research is required to explore the new approaches that could increase the productivity and lower the production cost. Nevertheless, biodiesel forms algae coupled with CO_2 sequestration, and wastewater recycling is the best alternative that will contribute to environment sustainability and future energy security.

References

Adeniyi OM, Azimov U, Burluka A (2018) Algae biofuel: current status and future applications. Renew Sustain Energy Rev 90:316–335

Arbib Z, Godos I, Ruiz J, Perales JA (2017) Optimization of pilot high rate algal ponds for simultaneous nutrient removal and lipids production. Sci Total Environ 589:66–72

Aresta M, Dibenedetto A, Carone M, Colonna T, Fagale C (2005) Production of biodiesel from macroalgae by supercritical CO_2 extraction and thermochemical liquifaction. Environ Chem Lett 3:136–139

Beal CM, Gerber LN, Sills DL, Huntley ME, Machesky SC, Walsh MJ et al. (2015) Algal biofuel production for fuels and feed in a 100-ha facility: a comprehensive technoeconomic analysis and life cycle assessment. Algal Res 10:266–279

Benemann JR, Oswald WJ (1996) Systems and economic analysis of microalgae ponds for conversion of CO_2 to biomass. Report number DOE/PC/93204/T5. Department of Civil Engineering, University of California, Berkley, CA, p 215

Bligh EG, Dyer WJ (1959) A rapid method of total lipid extraction and puri-cation. Can J Biochem Physiol 37:911–917

Brennan L, Owende P (2010) Biofuels from microalgae: a review of technologies for production, processing, and extractions of biofuels and co-products. Renew Sust Energ Rev 14:557–577

Brown AC, Knights BA, Conway E (1969) Hydrocarbon content and its relationship to physiological state in the green alga *Botryococcusbraunii*. Phytochemicals 8:543–547

Brown LM, Zeiler KG (1993) Aquatic biomass and carbon dioxide trapping. Energy Convers Manag 34:1005–1013

Campbell MN (2008) Biodiesel: algae as a renewable source for liquid fuel. Guelph Eng J 1:2–7

Carlsson AS, van Beilen JB, Moller R, Clayton D (2007) In: Bowles D (ed) Micro- and macro-algae: utility for industrial applications, outputs from the EPOBIO project. University of York, CPL Press, Newbury UK, p 86

Chakravarty S, Mallick N (2019) Optimization of lipid accumulation in an aboriginal green microalga *Selenastrum* sp. GA66 for biodiesel production. Biomass Bioenergy 126:1–1

Chen CY, Yeh KL, Aisyah R, Lee DJ, Chang JS (2011) Cultivation, photobioreactor design and harvesting of microalgae for biodiesel production: a critical review. Biores Technol 102:71–81

Chisti Y (2007) Biodiesel from microalgae. Biotechnol Adv 25:294–306

Chisti Y (2008) Biodiesel from microalgae beats bioethanol. Trends Biotechnol 26:126–131

Chisti Y, Yan J (2011) Energy from algae: current status and future trend algal biofuels—a status report. Appl Energy 88:3277–3279

Dayananda C, Sarada R, Usha Rani M, Shamala TR, Ravishankar GA (2007) Autotrophic cultivation of *Botryococcusbraunii* for the production of hydrocarbons and exopolysaccharides in various media. Biomass Bioenergy 31:87–93

Demirbas A (2008) Biofuels sources, biofuel policy, biofuel economy and global biofuel projections. Energy Convers Manage 49:2106–2116

Demirbas A, Demirbas MF (2011) Importance of algae oil as a source of biodiesel. Energy Convers Manage 52:163–170

Demirbas MF (2011) Biofuels from algae for sustainable development. Appl Energy 88:3473–3480

EIA (2016) International energy outlook 2016 with projections to 2040. Energy outlook 2016. U.S. Energy Information Administration, U.S. Department of Energy, Washington, DC

Feng FY, Yang W, Jiang GZ, Xu YN, Kuang TY (2005) Enhancement of fatty acid production of *Chlorella* sp. (Chlorophyceae) by addition of glucose and sodium thiosulphate to culture medium. Process Biochem 40:1315–1318

Feng P, Deng Z, Hu Z, Fan L (2011) Lipid accumulation and growth of *Chlorella zofingiensis* in flat plate photobioreactors outdoors. Biores Technol 102:10577–10584

Greenwell HC, Laurens LML, Shields RJ, Lovitt RW, Flynn KJ (2010) Placing microalgae on the biofuels priority list: a review of the technological challenges. J Am Oil Chem Soc 7:703–726

Grima ME, Belarbi EH, Fernandez FGA, Medina AR, Chisti Y (2003) Recovery of microalgal biomass and metabolites: process options and economics. Biotechnol Adv 20:491–515

Gul T (2016) Renewable transport fuel obligation statistics. Period 9 (2016/17): Department for Transport, p 1–6

Gutiérrez R, Gutiérrez-Sánchez R, Nafidi A (2008) Trend analysis using non homogeneous stochastic diffusion processes. Emission of CO_2; Kyoto protocol in Spain. Stoch Environ Res Risk Assess 22:57–66

Halim R, Danquah MK, Webley PA (2012) Extraction of oil from microalgae for biodiesel production: a review. Biotechnol Adv 30:709–732

Ho SH, Chen CY, Lee DJ, Chang JS (2011) Perspectives on microalgal CO_2 emission mitigation systems—a review. Biotechnol Adv 29:189–198
Ho SH, Chen WM, Chang JS (2010) *Scenedesmusobliquus* CNW-N as a potential candidate for CO_2 mitigation and biodiesel production. Biores Technol 101:8725–8730
Hsueh HT, Chu H, Yu ST (2007) A batch study on the bio-fixation of carbon dioxide in the absorbed solution from a chemical wet scrubber by hot spring and marine algae. Chemosphere 66:878–886
Huntley M, Redalje D (2007) CO_2 mitigation and renewable oil from photosynthetic microbes: a new appraisal. Mitig Adapt StratGl 12:573–608
Karpagam R, Preeti R, Ashokkumar B, Varalakshmi P (2015) Enhancement of lipid production and fatty acid profiling in *Chlamydomonasreinhardtii*, CC1010 for biodiesel production. Ecotoxicol Environ Saf 121:253–257
Kintisch E (2010) Climate change: senate looms as bigger hurdle after Copenhagen. Science 327:19–21
Krawczyk T (1996) Biodiesel alternative fuel makes inroads but hurdles remain. Inform 7:801–829
Kumar A, Guria C, Pathak AK (2018) Optimal cultivation towards enhanced algae biomass and lipid production using *Dunaliellatertiolecta* for biofuel application and potential CO_2 bio-fixation: effect of nitrogen deficient fertilizer, light intensity, salinity and carbon supply strategy. Energy 148:1069–1086
Li P, Miao X, Li R, Zhong J (2011) In situ biodiesel production from fast-growing and high oil content *Chlorella pyrenoidosa* in rice straw hydrolysate. J Biomed Biotechnol 7:1–8
Li T, Zheng Y, Yu L, Chen S (2013) High productivity cultivation of a heat-resistant microalga *Chlorella sorokiniana* for biofuel production. Biores Technol 131:60–67
Mallick N (2002) Biotechnological potential of immobilized algae for wastewater N, P and metal removal: a review. Biometals 15:377–390
Mallick N, Mandal S, Singh AK, Bishai M, Dash A (2012) Green microalga *Chlorella vulgaris* as a potential feedstock for biodiesel. J Chem Technol Biotechnol 87:137–145
Mata TM, Martins AA, Caetano NS (2010) Microalgae for biodiesel production and other applications: a review. Renew Sustain Energy Rev 14:217–232
Matsunaga T, Matsumoto M, Maeda Y, Sugiyama H, Sato R, Tanaka T (2009) Characterization of marine microalga, *Scenedesmus*sp. strain JPCC GA0024 toward biofuel production. Biotechnol Lett 31:1367–1372
McMillen S, Shaw P, Jolly N, Goulding B, Finkle V (2005) Biodiesel: fuel for thought, fuel for connecticut's future. Connecticut Center for Economic Analysis Report. University of Connecticut, p 56
Miao X, Wu Q (2004) High yield bio-oil production from fast pyrolysis by metabolic controlling of *Chlorella protothecoides*. J Biotechnol 110:85–93
Mohn FH (1980) Experiences and strategies in the recovery of biomass in mass culture of microalgae. In: Shelef G, Soeder CJ (eds) Algal biomass. Elsevier, Amsterdam, pp 547–571
Nevase SS, Gadge SR, Dubey AK, Kadu BD (2012) Economics of biodiesel production from Jatropha oil. J Agric Technol 8:657–662
Ota M, Takenaka M, Sato Y, Smith RL, Inomata H (2015) Variation of photoautotrophic fatty acid production from a highly CO_2 tolerant alga, *Chlorococcumlittorale*, with inorganic carbon over narrow ranges of pH. Biotechnol Prog 31:1053–1057
Petrusevski B, Bolier G, Van Breemen AN, Alaerts GJ (1995) Tangential flow filtration: a method to concentrate freshwater algae. Water Res 29:1419–1424
Piorreck M, Pohl P (1984) Formation of biomass, total protein, chlorophylls, lipids and fatty acids in green and blue-green algae during one growth phase. Phytochemicals 23:217–223
Rawat I, Ranjith KR, Mutanda T, Bux F (2013) Biodiesel from microalgae: a critical evaluation from laboratory to large scale production. Appl Energy 103:444–467
Rodolfi L, Zittelli GC, Bassi N, Padovani G, Biondi N, Bonini G (2008) Microalgae for oil: strain selection, induction of lipid synthesis and outdoor mass cultivation in a low-cost photobioreactor. Biotechnol Bioeng 102:100–112

Samorì C, Lopez Barreiro D, Vet R, Pezzolesi L, Brilman DWF, Galletti P, Tagliavini E. (2013) Effective lipid extraction from algae cultures using switchable solvents. Green Chem 15:353–356
Sheehan J, Dunahay T, Benemann J, Roessler P (1998) A look back at the US Department of Energy's aquatic species program: biodiesel from algae. NREL/TP-580–24190, National Renewable Energy Laboratory, USA
Slade R, Bauen A (2013) Micro-algae cultivation for biofuels: cost, energy balance, environmental impacts and future prospects. Biomass Bioenergy 53:29–38
Stewart C, Hessami MA (2005) A study of methods of carbon dioxide capture and sequestration the sustainability of a photosynthetic bioreactor approach. Energy Convers Manag 46:403–420
Ugwu CU, Aoyagi H, Uchiyama H (2008) Photobioreactors for mass cultivation of algae. Bioresour Technol 99:4021–4028
Ullah K, Ahmed M, Sofia Sharma VK, Lu P, Harvey A, Zafar M, Sultana S, Anyanwu CN (2014) Algal biomass as a global source of transport fuels: overview and development perspectives. Pro Nat Sci Mater 24:329–339
Viswanath B, Mutanda T, White S, Bux F (2010) The microalgae—a future source of biodiesel. Dyn Biochem Process Biotechnol Mol Biol 4:37–47
Wang B, Li Y, Wu N, Lan QC (2008) CO_2 bio-mitigation using microalgae. Appl Microbiol Biotechnol 79:707–718
Weissman JC, Goebel RP (1987) Design and analysis of microalgal open pond systems for the purpose of producing fuels: a subcontract report. Solar Energy Research Institute, US
Xu H, Miao X, Wu Q (2006) High quality biodiesel production from a microalgae *Chlorella protothecoides* by heterotrophic growth in fermenters. J Biotechnol 126:499–507
Yeh KL, Chang JS (2011) Nitrogen starvation strategies and photobioreactor design for enhancing lipid content and lipid production of a newly isolated microalga *Chlorella vulgaris* ESP-31: implications for biofuels. Biotechnol J 6:1358–1366
Zhang Q, Wang T, Hong Y (2014) Investigation of initial pH effects on growth of an oleaginous microalgae *Chlorella* sp. HQ for lipid production and nutrient uptake. Water Sci Technol 70:712–719

Ranjana Bhati is working as Assistant Professor at Department of microbiology, Bundelkhand University, Jhansi. She has completed her doctorate degree from Indian Institute of Technology, Kharagpur, in Algal Biotechnology. She has been awarded Dr. D. S. Kothari Postdoctoral Fellowship (UGC) at Jawaharlal Nehru University, New Delhi. Her area of research is Algal Biotechnology, Biofuels, Biopolymers, Applied and Environmental Microbiology. She has published several book chapters and research papers in referred to national and international journals.

Biotechnology of Biofuels: Historical Overview, Business Outlook and Future Perspectives

Vijay Lakshmi Jamwal, Nitika Kapoor and Sumit G. Gandhi

Abstract Continued reliance on fossil fuels is a major risk to energy security, and its extraction from natural reserves poses serious environmental threats. Exploration of new and renewable sources of energy is important to mitigate these concerns. In recent times, biofuels have come up as an inexhaustible and alternative source of energy which may be produced from plants or algae. Past 3–4 decades have seen a serious effort in this area and broadly can be classified in three generations of biofuels. Commonly cultivated crop plants served as an energy base for the production of the first generation of biofuel. However, this led to increase in food prices which had a serious negative impact on the third world countries. The second generation of biofuels was produced from non-crop high energy plants. However, both of these were dependent on agriculture, which requires high cultivable land reserves, considerable human resource involvement, irrigation facilities, is severely impacted by changes in rainfall patterns and weather conditions as well as biotic factors such as pathogen infections. In contrast, the third-generation biofuels are produced from microalgae which overcome the disadvantages of first- and second-generation biofuels. Microalgae are regarded to be an attractive source for energy due to its biomass productivity (dry weight per unit time per unit area) that is much higher than those of higher plants. Globally, several private industries and government organizations are involved in research and development of biofuels as alternative sources of energy. Governments around the world have provided subsidies in different forms for the production of biofuels. Brazil has come up as a world leader in sustainable biofuel production, with most automobiles running on bioethanol or bioethanol blends. Over the years, the use of biofuel has continuously increased in most countries. Recently, the increased pumping out of fossil fuels from Middle Eastern countries has led to a global decrease in crude oil putting pressure on alternative sources of fuels. However, this may not be sustainable in long term, and continued research on biofuels is necessary for future energy security. Present research trends include the isolation of new algae, with higher biomass production and oil accumulation. Further, transgenesis in algae has made it possible to increase photosynthetic rate, re-route metabolism

V. L. Jamwal · N. Kapoor · S. G. Gandhi (✉)
Plant Biotechnology Division, CSIR-Indian Institute of Integrative Medicine, Canal Road, Jammu 180001, India
e-mail: sumit@iiim.ac.in

A. Saxena (ed.), *Biotechnology Business - Concept to Delivery*, EcoProduction,
https://doi.org/10.1007/978-3-030-36130-3_6

toward higher oil production, make the algal membranes and cell walls withstand high oil, etc.

Keywords Biofuels · Fermentation · Fossil fuels · Metabolic engineering · Microalgae · Transesterification · Plant oil

1 Introduction

1.1 World Energy Requirement

Fossil fuels (coal, gas and oil) contribute more than 80% of the world's primary energy supply (Council WE 2016). These sources of energy are used for generating power, electricity, energy and heat for different purposes including industrial, domestic, commercial and transportation. Developing countries require to expand their economies with increase in population which results in voracious demand for fossil fuels. Greenhouse gases (CO_2) are released from the combustion of fossil fuels that cause global warming. According to International Energy Agency (IEA), there is a need to change the world's existing energy economy. It has been estimated that world's CO_2 emission will increase by 50% up to the year 2030. Nations are attempting to modify their existing energy mix and reduce reliance on fossil fuels so as to reduce the CO_2 emissions. International agreements by governments are turning to new forms of renewable energy, such as biomass, biofuels, geothermal, hydro, solar and wind, in order to reduce greenhouse gas emissions (Fig. 1). The IEA predicted in 2017 that the global capacity of renewable electricity will expand by over 920 GW between 2017 and 2022, with solar power comprising the largest component of the increase (Wold Energy Outlook 2017). The consumption of renewable energy sources for transportation is expected to elevate from over 4% in 2016 to 5% in 2022. Biofuels are expected to stand for over 90% of total renewable energy for transportation in 2022.

1.2 Conventional Sources of Energy

Conventional or non-renewable resources are not naturally replenished once they have been used. Non-renewable resources, such as fossil fuels, become economically inaccessible when they are excessively consumed (Khan 2006). The residues of living organisms were subject to high temperature and pressure over millions of years, which lead to the formation of fossil fuels. There are three types of fossil fuels used as source of energy: coal, oil and natural gas (Bertine and Goldberg 1971). Coal is a solid fossil fuel formed by the decomposition of plant remains over several thousands of years. Comparing the two types of fossil fuels, coal is relatively profuse. It is expected that the current coal supplies could last for more than 200 years. The

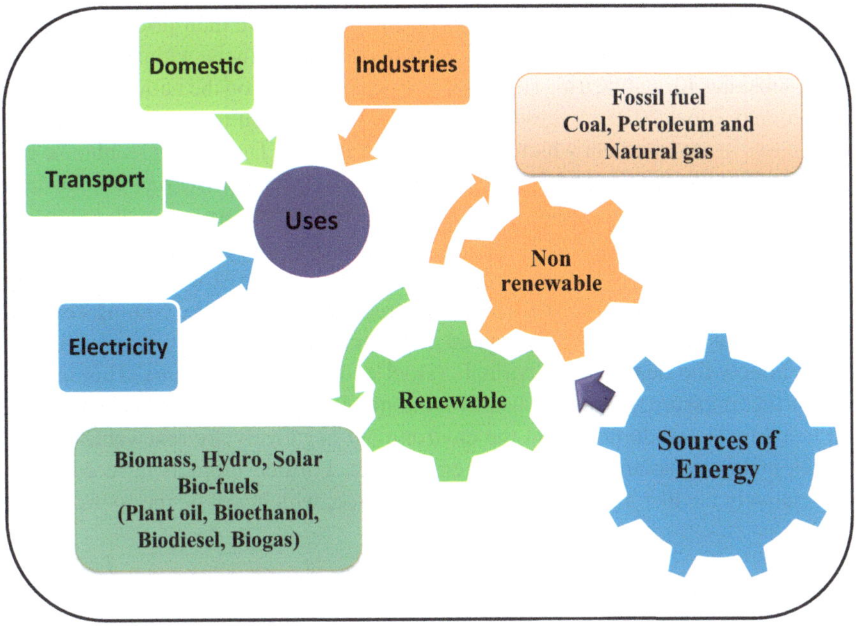

Fig. 1 Renewable and non-renewable sources of energy

consumption of coal has increased to double since the middle of 20th century. Many developing countries including China and India depend on coal for their energy supply (Bhattacharya and Jana 2009).

Petroleum oil is liquid fossil fuel that is formed from the decomposition of remains of animals and microorganisms under high pressure and high temperature. Oil is trapped in small spaces between the rocks and sediments. It is extracted by large drilling platforms (Vassoyevich 1967). Petroleum consists of mixture of various constituents such as petroleum gas, diesel, petrol and lubricating oil, which are separated by refining it (Meinschein 1959). Natural gas is a fossil fuel that is adaptable and comparatively cleaner than the other two forms of fossil fuel. In developed countries, use of natural gas has overtaken coal. Natural gas mainly consists of methane (CH_4). The reserves of natural gas are distributed equally over the globe as compared to the oil.

1.3 Depletion of Fossil Fuel Reserves and Environmental Impact

Growth in world's population and demand for urbanization, especially in underdeveloped and developing countries, has resulted in an extensive use of fossil fuels. The reckless wastage or suboptimal use of energy adds to the growing worry of fossil fuel extinction. With the rising fear of fossil fuels extinction, their prices have risen inexplicably. Several countries depend on importing fossil fuels, with some of them giving subsidies for public use. This puts excessive pressure on the nation's economy (Shafiee and Topal 2010). Excessive use of fossil fuels results in release of greenhouse gases in environment which cause global warming. The major greenhouse gas is carbon dioxide (CO_2), and 70–75% is emitted out from fossil fuel combustion (Davis and Caldeira 2010). These atmospheric changes in turn generate extensive climate changes globally. Melting of the world's glaciers and ice sheets due to rising temperatures is a major concern. This could lead to rise in sea levels, which could trigger significant population and infrastructure dislocations in the coming century. Further, use of fossil fuels is associated with increasing air pollution, which negatively affects the health of individuals (Hoel and Kverndokk 1996). This in turn adds further pressure on nation's health economy. Energy sources which can substitute fossil fuels and do not affect the environment negatively are renewable energy sources including solar energy, hydro energy, etc. Further, biofuels have come up as a relatively cleaner and renewable energy source, which will be the focus of this chapter.

1.4 Biofuels

Biofuels are fuels derived from biomass which comprises wood, agricultural crops and products, forest products, wastes and residues. Biofuels may be solid, liquid or gaseous. Solid biofuels include wood, charcoal, bagasse, etc. (Obernberger et al. 2006). Gaseous biofuels include methane gas which is produced from anaerobic fermentation of domestic waste, animal waste, wastewater treatment sludge, etc. (Stafford et al. 1980). Liquid biofuels include bioethanol, plant oils, vegetable oils and the methyl esters produced after transesterification of these oils which is commonly known as biodiesel (Granda et al. 2007).

2 Generation of Biofuels

Biofuels are classified in three different generations based on the source of feedstock (Fig. 2). Biofuels of first generation were primarily generated from food crops like grain, cane and plant oils, whereas biofuels of second generation were generated from energy crops other than food crops such as miscanthus, forest residues and woody biomass. In the third-generation biofuels, microalgae are used as source of biomass for biofuel production (Dutta et al. 2014).

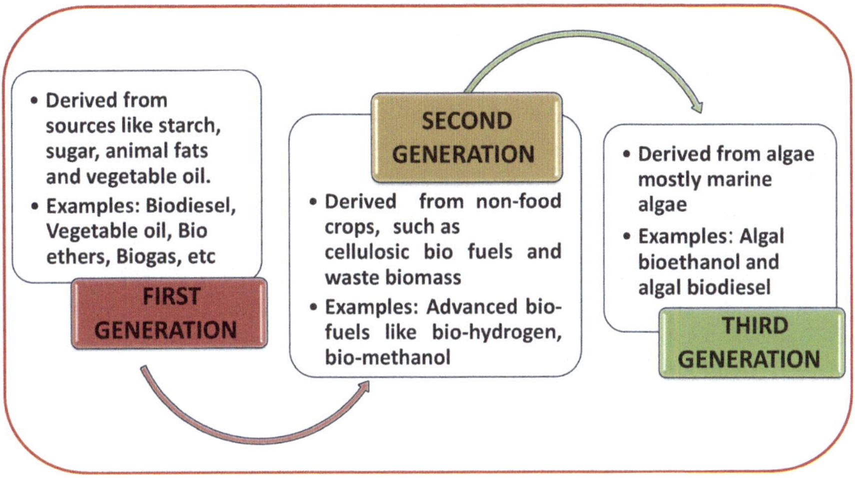

Fig. 2 Generations of biofuels

2.1 *First Generation*

2.1.1 Historical Overview

First-generation biofuels, also refer to conventional biofuels, are produced from sugar and starch from food crops or edible oil (Ho et al. 2014). First-generation biofuels are produced using different processes such as fermentation, distillation and transesterification (Ma and Hanna 1999). Sugars and starch are fermented to produce primarily ethanol, and in smaller quantities, butanol and propanol (Sarkar et al. 2012). Energy density of ethanol is less than gasoline, but it is presently used in various countries, including the USA, as an additive to gasoline (Hansen et al. 2005). Combustion of ethanol is comparatively cleaner than gasoline, thus emits out less greenhouse gases. Biodiesel is another first-generation biofuel which is produced through transesterification of plant or vegetable oils (Yusuf et al. 2011). Biodiesel can be used as alternative of petroleum diesel in diesel engines or by blending the two. These biofuels also supported agricultural industries and rural communities through increased demand for crops.

2.1.2 Sources

Plant oils as biodiesels: Rapeseed oil, sunflower oil, soya, castor, coconut oil, etc., are used as biodiesels. Transesterification of plant oils with alcohol results in fatty acid esters called fatty acid methyl ester (FAME) or fatty acid ethyl ester (FAEE), depending on the kind of alcohol used (Martín and Grossmann 2014). There are various physical and chemical properties of plant oils or vegetable oils that affect their suitability as fuels such as iodine value, ash content, cetane number, sulfur content and heat value. (Demirbas 2008). The ash, sulfur and potassium contents of the plant oil are some of the important characteristics. Plant oils or vegetable oils with low iodine value are more combustible and comparatively more efficient as fuel than oils with high iodine value. For example, coconut oil has low iodine value, whereas linseed oil has very high iodine value. Oils with low iodine value have high melting point and are solid at room temperature. Oils with low cetane numbers produce noise, cause trouble in starting and turn out thick exhaust smoke (e.g., linseed oil and rapeseed oil) (Schwab et al. 1987; Goering et al. 1982). The ash content of fuels is inversely proportional to the heating value. Biofuels have lower ash content and sulfur content as compared to the fossil fuels (Bozbas 2008).

Sugar-rich food crop for bioethanol: It is produced from the fermentation of different type of feedstocks that contain sugars or carbohydrates (Cardona et al. 2010). These feedstocks mainly consist of edible food crops such as barley, potato, rice, corn, sugarcane, wheat and vegetable oil such as mustard oil, olive oil, rapeseed oil, soybean oil, coconut oil, sunflower oil and canola oil. (Rulli et al. 2016). Bioethanol does not produce SO_2 or NO_x.

Biomass for biogas: It is formed by anaerobic decomposition of organic material such as biodegradable waste materials, including animal sewage and household waste. Under oxygen deprivation, anaerobic bacteria break down the organic matter to produce methane (Kigozi et al. 2013). Biogas consists of carbon dioxide, sulfur dioxide, methane and other gases in minute quantity. This method is used for sludge stability in wastewater treatment system. Small-scale and low-cost units for degradation of domestic waste are common in developing countries and production of gas for cooking (Zupančič and Grilc 2012).

2.1.3 Ecological and Environmental Impact

Biofuels emit less greenhouse gases as compared to the fossil fuels, thus positively affect the quality of air and water. First-generation biofuels included mainly ethanol (bioethanol) and biodiesel. Production of feedstocks for first-generation biofuels requires an agricultural bio-resource, considerable amount of water, fertilizers as well as pesticides (Escobar et al. 2009). Most of the environmental effects are related to the agricultural practices of the bio-energy crop cultivation. The net benefit of a biofuel is evaluated by analyzing its effect on air, water, soil, food and on net energy gain (Tilman et al. 2009). Biofuels derived from feedstocks produced from the plants grown on degraded non-cultivable lands, residues of crops, sustainably harvested forest remains as well as domestic and industrial waste (Escobar et al. 2009), will reduce their advantages though not competing with food crops, reducing effects on land clearing and providing real greenhouse gas reductions.

2.1.4 Economic Impact

A number of countries have experienced rapid boost in supply as well as demand of biofuels, but the US ethanol area distinctly stands out on the rate of growth experienced in the last ten years, which has made the USA the leading producer of biofuels worldwide (Gasparatos et al. 2013). In the USA, corn has been used for the production of ethanol for more than three decades. In 2011, the level of ethanol output increased by 80 folds relative to the levels in 1980. Methyl tertiary butyl ether (MTBE) was popularly used as oxygenate gasoline additive (Belpoggi et al. 1995). Ethanol has become known as the most feasible oxygenate alternative for MTBE, which also promotes a valuable market position for ethanol as a key additive of gasoline (Belpoggi et al. 1995).

2.1.5 Drawbacks and Need for Next Generation

First-generation biofuels have had severe limitations. The first key barrier is that the feedstocks they require are used for food. Across the world, a number of countries have cited biofuels as the reason for increased food prices, as supplies of crops for food have had to compete against crops used for fuel (Gabrielle 2008). Following, China's decision that food has priority (Duggan and Naarajärvi 2015), the USA also announced that it will be moving away from its corn ethanol fuel policy and into a new direction for biofuels.

The second key barrier is that certain first-generation biofuels have been estimated to have equal or higher CO_2 emissions than petrol when indirect land-use effects were taken into account. First-generation biofuels provide a small benefit over fossil fuels in regards to greenhouse gases which cause global warming, but cultivation and processing of feedstocks require very high amount of energy as compared to the energy supplied by them (Gabrielle 2008). Moreover, the agricultural intensification increases the adverse environmental effects including acidification, nutrient pollution, eco-toxicity and weakening of ozone cover (Doornbosch and Steenblik 2008; Quadrelli and Peterson 2007; Tomei and Upham 2009).

Mass production of feedstock for first-generation biofuels needs more cultivable lands which result in reduction of land for food crop production (Gabrielle 2008). Furthermore, the method for the production of feedstocks used for making first-generation biofuels is responsible for environmental deprivation. This decreased the enthusiasm about first-generation biofuels, thus potentiating the need for newer discoveries to mitigate the issues (Janda et al. 2012; Naqvi and Yan 2015).

2.2 *Second Generation*

2.2.1 Historical Overview

Biofuels of second-generation were designed to address the negative issues of first-generation biofuels that endangered food supply through the resources used. Production of second-generation biofuels comprises a range of non-food crops which includes energy crops, stalks of wheat and corn, woody trees, etc. (Eisentraut 2010). Second-generation biofuels mainly focused on two things, i.e., these biofuels utilize the non-food parts of food crops that are left over after the extraction of the food part. In these cases, the relevant crop parts are the stems, leaves and husks. Another focus has been to utilize non-food such as jatropha, palm and switchgrass. (Pimentel and Patzek 2005).

Second-generation biofuels also generate higher energy yields than first-generation fuels. The technology is fairly immature, so it still has potential of cost reductions and increased production efficiency as scientific advances occur (Antizar-Ladislao and Turrion-Gomez 2008). However, some biomasses used for the production of second-generation biofuels still compete with cultivable land used for food crops. In addition, the process to produce second-generation fuels is more elaborate, requires more energy and material than first-generation biofuels because it requires pre-treating the biomass to release the trapped sugars (Antizar-Ladislao and Turrion-Gomez 2008).

2.2.2 Sources

Different types of feedstocks are used for the production of the second generation of biofuels which includes energy crops such as amaranth, bamboo, eucalyptus grass, miscanthus, oilseed rape, poplar, salix, sugarbeet and sweet sorghum; agriculture and wood residues such as barn, citrus waste, corn stover, sugarcane bagasse, sawdust, wheat straw, waste rice-straw and wood; organic waste; vegetable oils such as jatropha oil and palm oil (Antizar-Ladislao and Turrion-Gomez 2008; Sanderson and Adler 2008). Cellulosic ethanol production requires freeing of sugar molecules from woody lignocellulosic material using enzymes, steam heating or other pre-treatments, followed by the fermentation of these sugars to produce bioethanol (Somerville 2007). Lignin is the resulting by-product which can be used as a fuel. The second-generation biofuel industry has had difficulty in producing ethanol from cellulose (Sims et al. 2010).

2.2.3 Ecological and Environmental Impact

Land usage: Feedstock in the second generation consists of switchgrass and miscanthus, which have relatively lower water and nutritional requirement and may positively affect the environment as these crops can grow on wastelands (Keshwani and Cheng 2009). Woody crops are also used as high energy feedstock. These have relatively high yield potential, are widely distributed geographically and require low levels of input in comparison with the annual crops (Smeets et al. 2007). In case of woody energy crops, use of sustainable forest practices can help in controlling deforestation (Ciccarese et al. 2012).

Air pollution: Second-generation biofuels emit relatively lesser amount of greenhouse gases as compared to fossil fuels, and they are in general more eco-friendly and socio-friendly as compared to first-generation biofuels (Dias et al. 2011). Woody energy crop obtained from forests stimulates indirect land-use change (iLUC) effects, which affect the balance of greenhouse gases (Witzke et al. 2010). Second-generation biofuels production by sustainably sourced wood (rather than fossil fuels) may reduce overall emissions.

Water usage: The characteristic feature of second-generation feedstocks (woody energy crop) includes higher transpiration rate due to their large root system, high leaf area index, long growing season and height (Versfeld and Van Wilgen 1986). High water use efficiency is important especially in the areas with limited water availability (Oliver et al. 2009).

Soil quality regulation: Second-generation feedstock (woody energy crop) can control the erosion of soil by wind and water also regulates the risk of flood. Trees require no annual tillage, provide soil cover throughout the year and exert a beneficial impact on soil properties including the enhancement of water fluxes leading to reduction in surface runoff, changes in size and stability of soil aggregates and a decrease in wind erosion (Kittredge 1948).

2.2.4 Economic Impact

First-generation biofuels contributed minimally to the mitigation of carbon emission and had upward effect on food prices as compared to the second generation (Mohr and Raman 2013). In addition to short-time gasolines pricing damping, trade improvements and wealth transfer, increased energy production in second-generation biofuels offered significant advantages. The use of cellulosic biomass for energy production resulted in significantly higher energy return on investment (EROI) and carbon sequestration in comparison with biofuels produced from starch and sugar (Tilman et al. 2006; Sheehan et al. 2003; Farrell et al. 2006).

The prices of second-generation biofuels depend significantly on the source of the feedstock, production process as well as its transport.

2.2.5 Drawbacks and Need for Next Generation

Second-generation biofuels continue harboring a number of limitations present in first-generation biofuels such as environmental pollution because of its combustion, water pollution because of the waste material from the processing and production of biofuels, and high land use. Further, unmanaged woody energy crops that are harvested unsustainably may significantly damage the environment and ecology due to deforestation.

The biofuels so far used have recorded limitations in speed of growth (in case of woody energy crop) (Kozlowski 1999) and yield compared (low yield in case of inedible crops) to the required volume of fuel, within a set timeframe. The production process for producing biofuels may be complex, multi-layered and time consuming. The land that can be used for the cultivation of food crops is limited, and harsher unused lands may sometimes not be utilized. Using land designated for food crops for the production of fuel would further reduce the land available for the cultivation of foods and put excessive pressure on food prices. All these factors make the production of second-generation biofuel unprofitable and commercially unsustainable. Microalgae are reasonable renewable source of energy as substitutes for first- and second-generation biofuels (Saladini et al. 2016). Microalgae can avail various types of renewable biofuels, such as biodiesel, bioethanol, methane and bio-hydrogen. Microalgae are capable of producing 15–300 times more biodiesel as compared to the traditional crop used for biofuel production (Lee and Lavoie 2013). The life cycle of microalgae is very short, and growth rate is very high. Additionally, the production of microalgae does not require high-quality cultivable land (Dragone et al. 2010).

2.3 *Third Generation*

2.3.1 Historical Overview

Microalgae are single-celled microorganisms which occur in freshwater and marine environment. Microalgae convert solar energy more efficiently as compared to higher plants. Due to high photosynthetic rate, microalgae are excellent candidates for fast and increased biomass production, lipid fabrication and biofuel production. In third-generation biofuels, microalgae biomass is used as feedstock for the production of biofuels. Microalgae are capable of growing throughout the year, and thus, oil yield per area of microalgae cultures could significantly enhance the yield of best oilseed crops. Cultivation of microalgae requires less water as compared to the food crops as well as energy crops and does not require herbicide or pesticide (Brennan and Owende 2010; Mata et al. 2010; Um and Kim 2009). Further, microalgae feedstock can grow on non-cultivable land, can use salty-marine water and thus does not compete with first- and second-generation feedstocks.

2.3.2 Sources

Microalgae are used as the major source for the production of third-generation biofuels. Microalgae are grown in a variety of aquatic environments, such as fresh or seawater, domestic or industrial wastewaters. Microalgae required sufficient quantity of carbon, nitrogen and phosphorus for their growth (Chen et al. 2011). Biofuel production by using microalgal biomass first includes the selection of microalgal strain, cultivation of microalgae, harvesting, drying of biomass and extraction of oil for production of bioethanol and biodiesel.

(a) **Cultivation of microalgae**

Cultivation of microalgae can be done by using two different cultivation systems: one is suspended cultures which include open ponds and closed-loop system (not exposed to air) or more complex systems such as photo-bioreactors (Chen et al. 2011).

(b) **Harvesting of microalgae**

Harvesting of microalgae has been thought-out as main blockage on the way to the commercial-scale processing of microalgae biomass for biofuel production. The cost of recovering the microalgal biomass from the medium is 20–30% of total biomass production cost (Grima et al. 2003). Harvesting of microalgal biomass can be attained using various chemical, physical and biological ways such as filtration, auto-flotation, air-flotation and flocculation. (Salim et al. 2011; Vandamme et al. 2013).

(c) **Extraction of oil algal biomass**

Drying of algae is required prior to lipid extraction. Sun drying the biomass is probably the cheapest method of drying, followed by cell disruption that includes high-pressure homogenisers, autoclaving and addition of hydrochloric acid, sodium hydroxide or alkaline lysis (Brennan and Owende 2010; Schenk et al. 2008; Mendes-Pinto et al. 2001). Several methods are used for extraction of lipids from microalgal biomass. Most commonly used methods are oil pressing, solvent extraction, supercritical fluid extraction (SFE) (Clifford and Williams 2000), etc.

After the extraction of microalgal oil, it is processed to convert into biodiesel through transesterification (Vasudevan and Briggs 2008). The transesterification reaction consists of transforming triglycerides into fatty acid alkyl esters, in the presence of an alcohol, such as methanol or ethanol, and a catalyst. Production of bioethanol from microalgae biomass is carried out through starch hydrolysis followed by fermentation and then distillation. It requires less energy and simplified processes as compared to the production of biodiesel (Harun et al. 2010).

2.3.3 Ecological and Environmental

Large-scale production of microalgae also has variety of environmental impacts. These environmental impacts include

(a) **Land use**

Production of microalgae can be done on marginalized land, which is unusable for the cultivation of crops; thereby, it minimizes the competition with food production (Borowitzka 1999). It requires considerably less land compared to that required for the cultivation of plants used for the production of first-generation or second-generation biofuels. However, topography and soil type may be a factor affecting the land availability for raceway pond systems as the installation of large shallow ponds requires relatively flat land. This may especially be a concern in hilly areas. Depending on the permeability of soil, there may be requirement of lining and sealing of pond for the cultivation of microalgae (Lundquist et al. 2010).

(b) **Nutrient pollution**

Leaching of residual nutrients culture medium into local aquatic systems causes negative as well as positive impact. Negative impact includes nutrient pollution (eutrophication), whereas a positive impact includes use of algae production in the treatment of water bodies which are suffering from excess nutrient supply (Graneli et al. 2008; AquaFUELs 2011).

(c) **Toxic effect of algae**

Algal species can produce toxins at certain stages of their life cycle. The effect of these toxins ranges from acute to chronic (Collins 1978). The production of toxins may depend on environmental conditions and the species and strain or algae that were used. Due to leaching of culture or the discarded wastewater from the pond, these toxins may be released in the water bodies and affect the environment (Collins 1978; Rellán et al. 2009).

(d) **Carbon fixation**

Carbon dioxide is required for the cultivation of microalgae, and theoretically, CO_2 fixation efficiency of 20–90% can be achieved. In open ponds, the efficiency of CO_2 fixation is less than 10%, whereas for thin layer cultivation, the CO_2 fixation efficiency is approximately 35%. CO_2 fixation efficiency of about 75% has been reported for closed photo-bioreactors (Weissman et al. 1988; Acién et al. 2012). Carbon neutral alternative renewable fuels are sought after for improvement of air quality and reducing the adverse effects on environment. Microalgae cultivation practices may be optimized to attain a high rate of carbon sequestration so that the amount of CO_2 released from the biofuel may almost equal the amount fixed during photosynthesis (Pokoo-Aikins et al. 2010).

2.3.4 Economic Impact

Japan pioneered large-scale commercial production of microalgae as an alternative food. (Krauss 1962). Several other countries followed the lead during 1970s and 1980s. By the year 2004, the industrial production of microalgae reached 7000 tonnes of dry mass per annum (Werner et al. 2004).

Economy has significant role in the commercial feasibility of production of biofuels using microalgae. Cost of microalgal oil production relies on different factors, such as biomass yield, oil content, scale of production systems and cost of recovering oil (Rodolfi et al. 2009). The factors that can sway the cost of production of microalgae include illumination, stirring, photosynthetic efficiency, medium for cultivation and CO_2. The production costs may further reduce, through optimization of these factors, making algal biofuel more luring and a viable alternative to fossil fuels, especially as the production of fossil fuels would decline in coming years due to the depletion of reserves. More than one type of biofuels can be extracted from algal biomass, which increases the value of biomass.

3 Business Outlook

3.1 Production

About 85% of world's ethanol production occurs in the USA and Brazil, followed by EU and China. According to Renewable Fuels Association (RFA), world's ethanol production improved from 2007 levels and reached their highest in 2017 after plunging briefly during the 2011 and 2012. In 2017, the USA attained a record production of 16 billion gallons of bioethanol (Association RF 2010).

According to OECD-FAO Agricultural Outlook 2017–2026, bioethanol and biodiesel production in developed countries will increase between 2017 and 2023, and then, it will slightly decrease after 2023. But in developing countries, the production of bioethanol is projected to keep increasing from 2017 to 2026. The trends of bioethanol production in world are predicted to increase by 11% and biodiesel by 8% between 2017 and 2026 (OECD-FAO 2017).

3.2 Consumption

According to IEA, globaly, the rate of production of biofuel is not able to keep up with increasing demand. In 2018, consumption of biofuels increased by 7% and reach to 152 billion liters, but average production increased only 3% per year. This will result in falling short sustained annual growth by 10% through to 2030.

Ethanol consumption worldwide is predicted to increase by about 12 billion liters and by the year 2026. Developing countries like Brazil, India, China and Thailand will account for 80% of the increased consumption. Use of ethanol may increase by 5.4 billion liters in Brazil alone. Consumption of ethanol is expected to increase by a billion liters in China as well (OECD-FAO 2018).

Over the next few years, biodiesel use will peak out in developed countries, while it may continue to increase firmly in developing countries. By 2027, Indonesia may annually consume 4.1 billion liters of biodiesel, whereas in Argentina and Brazil, its demand may grow up to 1.9 billion liters and 5.6 billion liters, respectively. Requirement of biodiesel will also increase in Colombia, Malaysia, Paraguay, India, Philippines and Thailand because of blending regulations. In most countries, currently, only 1–3% of total diesel requirement is fulfilled with biodiesel (OECD-FAO 2018).

3.3 Trade

Factors such as increasing domestic consumption within the countries that are presently exporting bioethanol, decreased demand from the countries that presently import bioethanol due to a boost in domestic production, as well as increasing reliance on alternate energy sources for transportation, such as electricity, will impact the global trade volumes of bioethanol, which is expected to marginally reduce by about 1% between 2017 and 2027. The USA may remain a net exporter of bioethanol produced from maize feedstock. Export of bioethanol from Brazil may slightly reduce due to increasing domestic demand. Import of bioethanol by countries like Japan and Canada is also expected to reduce (OECD-FAO 2018).

For similar reasons, as mentioned for biofuels, the trade volumes of biodiesel are also expected to decline, worldwide. Argentina is expected to remain as topmost exporter of biodiesel, followed by other countries such as Malaysia, Indonesia and Canada (OECD-FAO 2018).

4 Future Perspectives

First-generation biofuels have been phased out from most countries except in some developed nations, where corn starch is being used for production of bioethanol (Mohanty and Swain 2019). Second-generation biofuel, due to higher costs, is no longer a viable alternative. Third-generation biofuels are the major thrust area (Neto et al. 2019). Recently, there has been a decrease in crude oil (fossil fuel) prices internationally due to political issues and over-pumping in the Middle East nations (Bantacut et al. 2019). This has put excessive pressure on the biofuel industry, which was even earlier under economic stress. Moreover, the subsidies for biofuel industry have also been continuously declining, at least in the developed nations. For the third generation, microalgal biofuel to become economically viable and sustainable, its production capacity needs a major boost (Maity et al. 2014). There is a need for much higher biomass accumulation with significantly higher oil content, preferably in open ponds, throughout the year (Park et al. 2011). This presents major challenges because open ponds are subject to contamination from the environment (Day et al. 2012). Seasonal variations can have devastating effects on algal biomass (Olofsson et al. 2012). Water requirements are also an important issue which needs to be addressed as this also adds to the cost (Farooq et al. 2015). In addition, harvesting of small-sized microalgae from high volume liquid culture is energy and cost intensive. Compared to the conventional agriculture practice, cultivation of microalgae is more expensive, complex and requires specialized expertise, adding to the costs incurred on trained manpower (Vandamme et al. 2013).

Some of these difficulties may be overcome through biotechnological intervention. This may involve technological strategies such as: (a) development of biorefinery (Galkin and Ananikov 2019), (b) development of cost-effective technologies for biofuel production (He et al. 2018; Correa et al. 2019), (c) strain selection, such that it may be cultivated throughout the year without much impact of seasonal variation (Gnouma et al. 2018), (d) selected strains may be cultivable in saline seawater, with plants developed along the coast, to reduce the requirement of freshwater (Sydney et al. 2019) and (e) development of genetic tools and genome editing methodologies to engineer metabolism for higher biomass and lipid production (Naghshbandi et al. 2019; Jagadevan et al. 2018).

References

Acién F, Fernández J, Magán J, Molina E (2012) Production cost of a real microalgae production plant and strategies to reduce it. Biotechnol Adv 306:1344–1353

Antizar-Ladislao B, Turrion-Gomez JL (2008) Second-generation biofuels and local bioenergy systems. Biofuels Bioprod Bioref 25:455–469

AquaFUELs (2011) Report on biology and biotechnology of algae with indication of criteria for strain selection

Association RF (2010) Ethanol industry outlook. Renew Fuels Assoc 1–32

Bantacut T, Romli M, Noor E (2019) Life cycle assessment of cane-sugar in Indonesian sugar mill: energy use and GHG emissions. In: IOP conference series: materials science and engineering. IOP Publishing, p 012059

Belpoggi F, Soffritti M, Maltoni C (1995) Methyl-tertiary-butyl ether (MTBE)—a gasoline additive—causes testicular and lympho haematopoietic cancers in rats. Toxicol Ind Health 112:119–149

Bertine K, Goldberg ED (1971) Fossil fuel combustion and the major sedimentary cycle. Science 1733993:233–235

Bhattacharya S, Jana C (2009) Renewable energy in India: historical developments and prospects. Energy 348:981–991

Borowitzka MA (1999) Commercial production of microalgae: ponds, tanks, and fermenters. In: Progress in industrial microbiology, vol 35. Elsevier, pp 313–321

Bozbas K (2008) Biodiesel as an alternative motor fuel: production and policies in the European Union. Renew Sust Energy Rev 122:542–552

Brennan L, Owende P (2010) Biofuels from microalgae—a review of technologies for production, processing, and extractions of biofuels and co-products. Renew Sust Energy Rev 142:557–577

Cardona C, Quintero J, Paz I (2010) Production of bioethanol from sugarcane bagasse: status and perspectives. Biores Technol 10113:4754–4766

Chen C-Y, Yeh K-L, Aisyah R, Lee D-J, Chang J-S (2011) Cultivation, photobioreactor design and harvesting of microalgae for biodiesel production: a critical review. Biores Technol 1021:71–81

Ciccarese L, Mattsson A, Pettenella D (2012) Ecosystem services from forest restoration: thinking ahead. New Forest 435, 436:543–560

Clifford AA, Williams JR (2000) Introduction to supercritical fluids and their applications. In: Supercritical fluid methods and protocols. Springer, pp 1–16

Collins M (1978) Algal toxins. Microbiol Rev 424:725

Correa DF, Beyer HL, Possingham HP, Thomas-Hall SR, Schenk PM (2019) Global mapping of cost-effective microalgal biofuel production areas with minimal environmental impact Gcb. Bioenergy

Council WE (2016) World energy resources 2016 World Energy Council, London, UK

Davis SJ, Caldeira K (2010) Consumption-based accounting of CO_2 emissions. In: Proceedings of the national academy of sciences, vol 10712, pp 5687–5692

Day JG, Slocombe SP, Stanley MS (2012) Overcoming biological constraints to enable the exploitation of microalgae for biofuels. Biores Technol 109:245–251

Demirbas A (2008) Relationships derived from physical properties of vegetable oil and biodiesel fuels. Fuel 878–9:1743–1748

Dias MO, da Cunha MP, Maciel Filho R, Bonomi A, Jesus CD, Rossell CE (2011) Simulation of integrated first and second generation bioethanol production from sugarcane: comparison between different biomass pretreatment methods. J Ind Microbiol Biotechnol 388:955–966

Doornbosch R, Steenblik R (2008) Biofuels: is the cure worse than the disease? Revista Virtual REDESMA 2:63

Dragone G, Fernandes BD, Vicente AA, Teixeira JA (2010) Third generation biofuels from microalgae

Duggan N, Naarajärvi T (2015) China in global food security governance. J Contemp China 2495:943–960

Dutta K, Daverey A, Lin J-G (2014) Evolution retrospective for alternative fuels: first to fourth generation. Renew Energy 69:114–122

Eisentraut A (2010) Sustainable production of second-generation biofuels

Escobar JC, Lora ES, Venturini OJ, Yáñez EE, Castillo EF, Almazan O (2009) Biofuels: environment, technology and food security. Renew Sust Energy Rev 136–7:1275–1287

Farooq W, Suh WI, Park MS, Yang J-W (2015) Water use and its recycling in microalgae cultivation for biofuel application. Biores Technol 184:73–81

Farrell AE, Plevin RJ, Turner BT, Jones AD, O'hare M, Kammen DM (2006) Ethanol can contribute to energy and environmental goals. Science 3115760:506–508

Gabrielle B (2008) Significance and limitations of first generation biofuels. Journal de la Société de biologie 2023:161–165

Galkin KI, Ananikov VP (2019) Towards improved biorefinery technologies: 5-methylfurfural as a versatile C6 platform for biofuels development. ChemSusChem 121:185–189

Gasparatos A, Stromberg P, Takeuchi K (2013) Sustainability impacts of first-generation biofuels. Animal Front 32:12–26

Gnouma A, Sehli E, Medhioub W, Dhieb RB, Masri M, Mehlmer N, Slimani W, Sebai K, Zouari A, Brück T (2018) Strain selection of microalgae isolated from Tunisian coast: characterization of the lipid profile for potential biodiesel production. Bioprocess Biosyst Eng 4110:1449–1459

Goering C, Schwab A, Daugherty M, Pryde E, Heakin A (1982) Fuel properties of eleven vegetable oils. Trans ASAE 256:1472–1477

Granda CB, Zhu L, Holtzapple MT (2007) Sustainable liquid biofuels and their environmental impact. Environ Prog 263:233–250

Graneli E, Weberg M, Salomon PS (2008) Harmful algal blooms of allelopathic microalgal species: the role of eutrophication. Harmful Algae 81:94–102

Grima EM, Belarbi E-H, Fernández FA, Medina AR, Chisti Y (2003) Recovery of microalgal biomass and metabolites: process options and economics. Biotechnol Adv 207–8:491–515

Hansen AC, Zhang Q, Lyne PW (2005) Ethanol–diesel fuel blends—a review. Biores Technol 963:277–285

Harun R, Singh M, Forde GM, Danquah MK (2010) Bioprocess engineering of microalgae to produce a variety of consumer products. Renew Sust Energy Rev 143:1037–1047

He Y, Wu T, Wang X, Chen B, Chen F (2018) Cost-effective biodiesel production from wet microalgal biomass by a novel two-step enzymatic process. Biores Technol 268:583–591
Ho DP, Ngo HH, Guo W (2014) A mini review on renewable sources for biofuel. Biores Technol 169:742–749
Hoel M, Kverndokk S (1996) Depletion of fossil fuels and the impacts of global warming. Res Energy Econ 182:115–136
Jagadevan S, Banerjee A, Banerjee C, Guria C, Tiwari R, Baweja M, Shukla P (2018) Recent developments in synthetic biology and metabolic engineering in microalgae towards biofuel production. Biotechnol Biofuels 111:185
Janda K, Kristoufek L, Zilberman D (2012) Biofuels: policies and impacts. Agric Econ 588:372–386
Keshwani DR, Cheng JJ (2009) Switchgrass for bioethanol and other value-added applications: a review. Biores Technol 1004:1515–1523
Khan B (2006) Non-conventional energy resources. Tata McGraw-Hill Education
Kigozi R, Aboyade A, Muzenda E (2013) Biogas production using the organic fraction of municipal solid waste as feedstock
Kittredge J (1948) Forest influences: the effects of woody vegetation on climate, water, and soil, with applications to the conservation of water and the control of floods and erosion forest influences: the effects of woody vegetation on climate, water, and soil, with applications to the conservation of water and the control of floods and erosion
Kozlowski T (1999) Soil compaction and growth of woody plants. Scand J For Res 146:596–619
Krauss RW (1962) Mass culture of algae for food and other organic compounds. Am J Bot 494:425–435
Lee RA, Lavoie J-M (2013) From first-to third-generation biofuels: challenges of producing a commodity from a biomass of increasing complexity Animal. Frontiers 32:6–11
Lundquist TJ, Woertz IC, Quinn N, Benemann JR (2010) A realistic technology and engineering assessment of algae biofuel production. Energy Biosci Inst 1
Ma F, Hanna MA (1999) Biodiesel production: a review. Biores Technol 701:1–15
Maity JP, Bundschuh J, Chen C-Y, Bhattacharya P (2014) Microalgae for third generation biofuel production, mitigation of greenhouse gas emissions and wastewater treatment: present and future perspectives—a mini review. Energy 78:104–113
Martín M, Grossmann IE (2014) Simultaneous optimization and heat integration for the coproduction of diesel substitutes: biodiesel (FAME and FAEE) and glycerol ethers from algae oil. Ind Eng Chem Res 5328:11371–11383
Mata TM, Martins AA, Caetano NS (2010) Microalgae for biodiesel production and other applications: a review. Renew Sust Energy Rev 141:217–232
Meinschein W (1959) Origin of petroleum. AAPG Bulletin 435:925–943
Mendes-Pinto M, Raposo M, Bowen J, Young A, Morais R (2001) Evaluation of different cell disruption processes on encysted cells of Haematococcus pluvialis: effects on astaxanthin recovery and implications for bio-availability. J Appl Phycol 131:19–24
Mohanty SK, Swain MR (2019) Bioethanol production from corn and wheat: food, fuel, and future. In: Bioethanol production from food crops. Elsevier, pp 45–59
Mohr A, Raman S (2013) Lessons from first generation biofuels and implications for the sustainability appraisal of second generation biofuels. Energy Policy 63:114–122
Naghshbandi MP, Tabatabaei M, Aghbashlo M, Aftab MN, Iqbal I (2019) Metabolic engineering of microalgae for biofuel production
Naqvi M, Yan J (2015) First-generation biofuels. Handbook Clean Energy Syst 1–18
Neto JM, Komesu A, da Silva Martins LH, Gonçalves VO, de Oliveira JAR, Rai M (2019) Third generation biofuels: an overview. In: Sustainable bioenergy. Elsevier, pp 283–298
Obernberger I, Brunner T, Bärnthaler G (2006) Chemical properties of solid biofuels—significance and impact. Biomass Bioenergy 3011:973–982
Oliver RJ, Finch JW, Taylor G (2009) Second generation bioenergy crops and climate change: a review of the effects of elevated atmospheric CO_2 and drought on water use and the implications for yield Gcb. Bioenergy 12:97–114

Olofsson M, Lamela T, Nilsson E, Bergé JP, Del Pino V, Uronen P, Legrand C (2012) Seasonal variation of lipids and fatty acids of the microalgae Nannochloropsis oculata grown in outdoor large-scale photobioreactors. Energies 55:1577–1592

Park J, Craggs R, Shilton A (2011) Wastewater treatment high rate algal ponds for biofuel production. Biores Technol 1021:35–42

Pimentel D, Patzek TW (2005) Ethanol production using corn, switchgrass, and wood; biodiesel production using soybean and sunflower. Nat Res Res 141:65–76

Pokoo-Aikins G, Nadim A, El-Halwagi MM, Mahalec V (2010) Design and analysis of biodiesel production from algae grown through carbon sequestration. Clean Technol Environ Policy 123:239–254

Publishing O, Food, Organization A (2017) OECD-FAO agricultural outlook 2017–2026, OECD Publishing

Publishing O, Food, Organization A (2018) OECD-FAO agricultural outlook 2018–2027, OECD Publishing

Quadrelli R, Peterson S (2007) The energy–climate challenge: recent trends in CO_2 emissions from fuel combustion. Energy Policy 3511:5938–5952

Rellán S, Osswald J, Saker M, Gago-Martinez A, Vasconcelos V (2009) First detection of anatoxin-a in human and animal dietary supplements containing cyanobacteria. Food Chem Toxicol 479:2189–2195

Rodolfi L, Chini Zittelli G, Bassi N, Padovani G, Biondi N, Bonini G, Tredici MR (2009) Microalgae for oil: strain selection, induction of lipid synthesis and outdoor mass cultivation in a low-cost photobioreactor. Biotechnol Bioeng 1021:100–112

Rulli MC, Bellomi D, Cazzoli A, De Carolis G, D'Odorico P (2016) The water-land-food nexus of first-generation biofuels. Sci Rep 6:22521

Saladini F, Patrizi N, Pulselli FM, Marchettini N, Bastianoni S (2016) Guidelines for energy evaluation of first, second and third generation biofuels. Renew Sust Energy Rev 66:221–227

Salim S, Bosma R, Vermuë MH, Wijffels RH (2011) Harvesting of microalgae by bio-flocculation. J Appl Phycol 235:849–855

Sanderson M, Adler P (2008) Perennial forages as second generation bioenergy crops. Int J Mol Sci 95:768–788

Sarkar N, Ghosh SK, Bannerjee S, Aikat K (2012) Bioethanol production from agricultural wastes: an overview. Renew Energy 371:19–27

Schenk PM, Thomas-Hall SR, Stephens E, Marx UC, Mussgnug JH, Posten C, Kruse O, Hankamer B (2008) Second generation biofuels: high-efficiency microalgae for biodiesel production. Bioenergy Res 11:20–43

Schwab A, Bagby M, Freedman B (1987) Preparation and properties of diesel fuels from vegetable oils. Fuel 6610:1372–1378

Shafiee S, Topal E (2010) A long-term view of worldwide fossil fuel prices. Appl Energy 873:988–1000

Sheehan J, Aden A, Paustian K, Killian K, Brenner J, Walsh M, Nelson R (2003) Energy and environmental aspects of using corn stover for fuel ethanol. J Ind Ecol 73–4:117–146

Sims RE, Mabee W, Saddler JN, Taylor M (2010) An overview of second generation biofuel technologies. Biores Technol 1016:1570–1580

Smeets E, Bouwman A, Stehfest E (2007) Interactive comment on "N_2O release from agro-biofuel production negates global warming reduction by replacing fossil fuels" by Crutzen PJ et al. Atmos Chem Phys Discuss 7:S4937–4941

Somerville C (2007) Biofuels Curr Biol 174:R115–R119

Stafford DA, Hawkes DL, Horton R (1980) Methane production from waste organic matter

Sydney EB, Sydney ACN, de Carvalho JC, Soccol CR (2019) Microalgal strain selection for biofuel production. In: Biofuels from algae. Elsevier, pp 51–66

Tilman D, Hill J, Lehman C (2006) Carbon-negative biofuels from low-input high-diversity grassland biomass. Science 3145805:1598–1600

Tilman D, Socolow R, Foley JA, Hill J, Larson E, Lynd L, Pacala S, Reilly J, Searchinger T, Somerville C (2009) Beneficial biofuels—the food, energy, and environment trilemma. Science 3255938:270–271
Tomei J, Upham P (2009) Argentinean soy-based biodiesel: an introduction to production and impacts. Energy Policy 3710:3890–3898
Um B-H, Kim Y-S (2009) A chance for Korea to advance algal-biodiesel technology. J Ind Eng Chem 151:1–7
Vandamme D, Foubert I, Muylaert K (2013) Flocculation as a low-cost method for harvesting microalgae for bulk biomass production. Trends Biotechnol 314:233–239
Vassoyevich N (1967) Theory of the sedimentary-migration origin of oil: an SSSR Izvestiya, ser. geol., no. 11
Vasudevan PT, Briggs M (2008) Biodiesel production—current state of the art and challenges. J Ind Microbiol Biotechnol 355:421
Versfeld D, Van Wilgen B (1986) Impact of woody aliens on ecosystem properties. In: The ecology and management of biological invasions in Southern Africa. Proceedings of the national synthesis symposium on the ecology of biological invasions. Oxford University Press, pp 239–246
Weissman JC, Goebel RP, Benemann JR (1988) Photobioreactor design: mixing, carbon utilization, and oxygen accumulation. Biotechnol Bioeng 314:336–344
Werner A, Clarke D, Kraan S (2004) Strategic review and the feasibility of seaweed aquaculture in Ireland, Marine Institute
Witzke HP, Fabiosa JF, Gay SH, Golub AA, Havlik P, Msangi S, Tokgoz S, Searchinger TD (2010) A decomposition approach to assess iLUC results from global modeling efforts
Wold Energy Outlook (2017)
Yusuf N, Kamarudin SK, Yaakub Z (2011) Overview on the current trends in biodiesel production. Energy Conv Manage 527:2741–2751
Zupančič GD, Grilc V (2012) Anaerobic treatment and biogas production from organic waste. Manage Org Waste 3:1–28

Vijay Lakshmi Jamwal Masters in Human Genetics. Presently pursuing Ph.D. at AcSIR. Working at CSIR-Indian Institute of Integrative Medicine (Jammu), toward understanding the physiological and ecological roles of secondary metabolites in high value medicinal plants. Trained in molecular tools such as CRISPR-Cas9, transcriptomics, metagenomics, metabolomics and transgenics.

Nitika Kapoor Masters in Human Genetics. Awardee of ICMR Junior and Senior research fellowships. Presently, pursuing Ph.D. at CSIR-Indian Institute of Integrative Medicine, Jammu. Working on regulatory aspects of secondary metabolism in medicinal plants, employing cutting edge molecular tools, transcriptomics, metabolomics, etc.

Sumit G. Gandhi Working as senior faculty (Principal Scientist) at CSIR-IIIM (Jammu), on various aspects of secondary metabolism in microorganisms and medicinal & aromatic plants, since more than a decade. Ph.D. (JNU, Delhi) at CCMB (Hyderabad) focused on understanding molecular effectors and mechanisms of epigenetic regulation. Bachelors & Masters in Microbiology. Dr. Gandhi has resourcefully collaborated with industries and academia, on various projects and has been a recipient of grants from public and private sectors.

Regulations

Regulations for Health Care Biotechnology Products in Major Markets of the World

Durga Prasad Mindala, Yashbir S. Bedi, Satish Kumar Gupta, Sumit G. Gandhi and Inshad Ali Khan

Abstract Biotechnology products have applications in many areas such as medicine, agriculture, energy, and environmental protection. Drugs and food supplements are amongst the major biotechnology-based products that cater to the health care sector. The products that are intended for use in diagnosing, curing, mitigating, or treating disease fall in the category of a drug and regulations are applicable to it. As a general rule of thumb, such products must subscribe to the best available science at the time. Regulatory authorities/systems are entrusted with the task of securing health of common people, as well as safety of subjects and environment while accelerating economic growth, innovation, competitiveness, and job creation. The regulatory agency has the responsibility to review laboratory and clinical research before approval of such products for commercial use. Similarly, products such as food supplements that are intended to prevent a disease or help in ameliorating a disease condition also require appropriate regulations. Although the general principles on which the regulatory guidelines are based remain almost the same, minor procedural and nomenclature differences exist in the regulatory agencies in various countries. In the present article, we provide specific and compiled knowledge of current regulatory procedures in the European Union (EU), USA, and India.

Keywords Drug · Regulatory authorities · Biotechnology · European Union · USA and India

Abbreviations

ADME	Absorption, Distribution, Metabolism, and Excretion
BLA	Biologics License Application
CBER	Center for Biologics Evaluation and Research

D. P. Mindala · Y. S. Bedi · S. K. Gupta · S. G. Gandhi (✉) · I. A. Khan (✉)
Indian Institute of Integrative Medicine, Canal Road, Jammu 180001, India
e-mail: sumit@iiim.ac.in

I. A. Khan
e-mail: iakhan@iiim.ac.in

A. Saxena (ed.), *Biotechnology Business - Concept to Delivery*, EcoProduction,
https://doi.org/10.1007/978-3-030-36130-3_7

CDER	Center for Drug evaluation and Research
CNS	Central Nervous System
CVS	Cardiovascular System
DBT	Department of Biotechnology
DLC	District-Level Committees
eCTD	Electronic Common Technical Document
ETIPC	Emerging Technologies Interagency Policy Coordination Committee
EU	European Union
GEAC	Genetic Engineering and Approval Committee
GLP	Good Laboratory Practice
IAEC	Institutional Animal Ethics Committee
IB	Investigator's Brochure
ICH	International Conference on Harmonization
IDE	Investigational Device Exemption
INAD	Investigational New Animal Drug
IND	Investigational New Drug
IRB	Institutional Review Board
MoEF	Ministry of Environment and Forests
MS	Multiple Sclerosis
NDA	New Drug Application
NIH	National Institutes of Health
OECD	Organization for Economic Cooperation and Development
PD	Pharmacodynamics
PK	Pharmacokinetics
r-DNA	Recombinant Deoxy Ribonucleic Acid
RS	Respiratory System
SBCC	State Biotechnology Coordination Committees
SmPC	Summary of Product Characteristics
USDA	US Department of Agriculture
USEPA	US Environmental Protection Agency
USFDA	US Food and Drug Administration

1 Introduction

Biotechnology advancements have revolutionized the health care sector through biotechnology-derived pharmaceuticals (biological). Initial development of such products was reported in early 1980s; however, marketing permissions of these products took another 10 years. Regarding safety assessment of the biologics, several guidelines and documents were published from by various regulatory agencies at various time points. This information available with the regulatory authorities was of utmost use for the review/development of such products and their background in developing new biopharmaceuticals/biologics. A thorough analysis of this information is mandatory to develop scientifically acceptable preclinical safety evaluation programs.

"The Coordinated Framework for Regulation of Biotechnology" ref was finalized in 1986, after successful use of r-DNA techniques, and some of the formulations from such techniques were put to use commercially. "The Office of Science and Technology Policy", USA, was of the opinion that "existing statutes provide a basic network of agency jurisdiction over both research and products; this network forms the basis of this coordinated framework and helps assure reasonable safeguards for the public (OSTP 1986)". The existing laws passed by US legislative bodies on the subject were regulated by "US Environmental Protection Agency (EPA), the US, Food and Drug Administration (FDA), and the US Department of Agriculture (USDA)".

The stimulus for regulations on research and development of the r-DNA technology was provided in an International Conference held at Asilomar, California, in 1975. With passage of time, the onus for these regulations shifted from national to international regulatory authorities, resulting in the publication of the "blue book" titled, "***Recombinant DNA Safety Considerations***", by the Organization for Economic Cooperation and Development (OECD) (Development 1986). This blue book describes three general approaches to be considered while r-DNA organisms and r-DNA techniques. First, the risks rose by r-DNA organisms and conventional organisms are similar in nature. Second, the traits exhibited by genetically modified organisms through r-DNA techniques are more predictable in nature than regular methods of living cell manipulation. Third, legislation cannot be justified on scientific basis.

In addition, OECD and the National Institutes of Health (NIH) have made many other recommendations. As on day, countries at large observed the guidelines with different approaches, i.e. no regulation (south pacific countries), appropriate regulation (USA and some European countries), or stringent biotechnology specific regulation (German) (Brown 1991). Two distinct but closely interrelated regulatory agencies carry out regulatory functions to protect public health and the safety of the public at large regulated by the government of the said country. These are consumer and occupational safety to consume biotechnology products and to enter a workplace where biotechnology products or biotechnological means of production are in use.

2 Classifications

Biotechnology and life sciences contribute to the modernization of industry and shall be classified into three broad groups:

a. Health care applications:
   It includes the development of advanced medicines, therapies, diagnostics, and vaccines. The generation of novel drugs for diseases, that did not exist earlier and upgradation of the existing therapeutics.

b. Agriculture, livestock, veterinary products, and aquaculture:
   In this sector, applications include the production of vaccines, improvement of crop production and animal feed, bettering the quality of livestock, focussed breeding, efficient food processing, and improved diagnostics for detecting a spectrum of diseases.
c. Industrial processes and manufacturing:
   Industrial applications include the use of enzymes in the production of various products of commercial use such as pulp and paper, detergents, textiles, and biomass. Most contemporary processes include fermentation and enzyme biocatalysis instead of traditional chemical synthesis.

3 Regulations

Regulatory authorities in the world aims providing safe items for its consumption and reliable regulatory systems. However, regulatory approach differs country wise with respect to manufacturing and commercialization of the biologics. Taking into consideration the set up in USA, the new Biotechnology Working Group under the "Emerging Technologies Interagency Policy Coordination Committee (ETIPC)" from the Executive Office of the President is governed as detailed at introduction section (Gottlieb 2018). The working group shall harmonize with other federal agencies and offices where ever needed. In Europe, national law implements legal measures through regulations that are immediately applicable. The legal approach provides necessary framework for the processes and/or products, including technical specifications enforced by the "European Committee for Standardization (CEN and CENELEC)", with the help of the national entities (90/220/EEC, E.C. 1990). These standards are adopted on a voluntary basis without any regulatory requirements. In India, "Department of Biotechnology (DBT)" constituted under the Ministry of Science and Technology is the nodal agency for policy, promotion of R&D, international cooperation, manufacturing activities, etc. With coordination of Genetic Engineering and Approval Committee (GEAC) constituted under "Ministry of Environment and Forests (MoEF)", CDSCO and DBT are the regulatory bodies in the area of biotechnology in India (Rao 2018).

3.1 In USA

FDA approves any new drug for its manufacture as well as commercialization only if its safety and effectiveness are proven through trials. The financial burden of these testing and trials is borne by the manufacturer/sponsor. The safety studies shall be conducted step wise as per the regulatory guidelines.

3.1.1 Pre-market Approval of the Biological Product

In the USA, "biological products" are licensed by Center for Biologics Evaluation and Research (CBER) thorough Biologics License application (BLA) process ensuring its safety, purity, and potency regulated by the laws in force in the said country. The nonbiological drug requires laboratory studies and animal testing to define their pharmacologic and toxicological effects before they can be studied in humans similar to that of the biological products under "Good Laboratory Practice (GLP) regulations". However, biologics require a "flexible, case-by-case, science-based approach" to preclinical testing.

Preclinical studies are undertaken to establish initial safe dose in human, potential target organs for toxicity, and safety parameters to be considered during clinical monitoring.

For biotechnology-derived pharmaceuticals, the FDA follows the "International Conference on Harmonization of Technical Requirements for the Registration of Pharmaceuticals for Human Use (ICH) S6 guidelines".

Sponsor must perform in vitro and in vivo assay studies to determine the pharmacological activity, i.e. pharamacokinetic (PK), pharmacodynamics (PD), and mechanism of action of the investigative agent (Group IEW 2011). Evaluate effectiveness/safety of the agent on specific organs and systems on these systems separately, like CNS, CVS, RS, etc., (Guideline IHT 2011). Sponsor shall also conduct dose-dependent biopharmaceutical studies like PK and toxicokinetic studies to measure absorption, disposition, and metabolism (ADME) with reference to antibody-mediated clearance and investigate dose-response relationships (Guideline IHT 2011). This information helps to determine safe dose for human to conduct the clinical studies. Immunogenicity and carcinogenicity testing might be included in the preclinical studies of the agent though animals are not always a good predictor of human immunogenecity (Guideline IHT 2011). Carcinogenicity studies may be done based on the "duration of clinical dosing, population, and biological activity (Guideline IHT 2011)". These studies are done as per ICH S6 guidelines. Similar to carcinogenicity studies, reproductive and developmental toxicity studies may or may not be required but based on "the product, clinical indication, and intended patient population".

3.1.2 Investigational New Drug Application (IND)

According to current regulations, new R&D products in the USA, which are not approved for marketing, need to be submitted as an IND to the office of Food and Drugs Administration (FDA). In USA, different sections/offices for the approval/licensing of different products are an official process, for example, nonbiological application shall be submitted to the "Center for Drug Evaluation and Research (CDER) and for biological Center for Biologics Evaluation and Research (CBER)" (Table 1). As per US regulations, 21 CFR 312.22 and 312.23 contains the

Table 1 Summary of IND application submission in regulatory market

Contents	United States	Europe	India
1. IND application submission/departments	Division of communication and consumer affairs, office of communication, outreach and development, center for biologics evaluation and research (CBER), food and drug administration (FDA)—10903	Heads of medical agency (HMA), European medicine agency (EMA), Spark building Orlyplein 24 1043 DP Amsterdam The Netherlands	Central drugs standard control organization (CDSCO) Directorate general of health services Ministry of health and family welfare Government of India
2. Mode of submission	Electronic common technical document (eCTD)	Electronic common technical document eCTD through EudraLink	Online submission through SUGUM online portal
3. Application forms	Cover sheet (Form FDA-1571)	Annex-1: clinical trial application form	Form CT-04
4. Application fee	$2,588, 477	EUR 87600 for initial request	INR 3,00,000
5. Guidelines	ICH S6, OECD and USFDA	ICH S6, OECD and EMA guidelines	OECD/ICH and schedule Y of drugs and cosmetic act, 1940
6. Directives	21 CFR 312.22, 312.23 and 312.30	Article 9(8) of directive 2001/20/EC	Schedule Y of drugs and cosmetic act, 1940

basics and the general requirements for an IND Submission (Regulations, C. o. F., 21 2018). An enumerated list of studies and documents are mentioned below.

3.1.3 Cover Sheet (Form FDA-1571) (CFR 2019)

Form FDA-1571 is an application form to seek the permission to conduct the clinical trials on the human subject. The application form contains the general information of the sponsor and investigational agent (Drug) (Table 1). Sponsor shall mention clearly in the application, the phase to be conducted. An "Institutional Review Board" (IRB) is constituted in every organization committed to comply with the requirements described in part 56 and is responsible for the approval and review of each of the studies in the proposed clinical investigation and report to the IRB (CFR, Part 56, Institutional Review Board 2018). Complete details of the persons responsible for monitoring the progress of the investigations as well as reviewing and evaluation of information relevant to the safety of the drug, has to be provided in the application.

3.1.4 Introductory Statement and General Investigational Plan

The Introductory statement should contain detailed information of the Active Pharmaceutical Ingredients (API) and the structural elucidation of the molecule, information on the formulation, its dosage form, route of administration, and the broad objectives and duration of the proposed clinical investigation(s). In addition to that, previous human experience with the drug, marketing experience in other countries that may be relevant to the safety of the drug (Regulations, C. o. F., 21 2018).

The investigational plan requires additional information such as

(a) The rationality for the drug, the research, and development study;
(b) The therapeutic indications;
(c) The basics to be followed in evaluating the drug;
(d) The kind of clinical trials to be conducted to the estimated number of patients; and
(e) Any risks of particular severity or seriousness anticipated on the basis of the preclinical studies or previous studies in humans with the drug or related drugs.

3.1.5 Investigator's Brochure (IB)

Investigator's brochure, containing the following information: An introduction on the drug substance and the formulation, along with the structural formula (structural elucidation of the drug molecule). The data on summary of the preclinical studies like toxicity, efficacy, and ADME of the drug in animals.

3.1.6 Investigational Protocols

Investigational protocols should be submitted in accordance with 312.30(a) (Regulations, C. o. F., 21 2018). The Phase 1 studies are primarily done to access the safety of the new product in humans. Phase 1 protocols give an outline of the number of human subjects to be enrolled, a description of safety and dosing plan including duration, dose, or method be used in quantify dose. The details of the paramcters critical for the safety arc given in detail. Phase 2 and 3 studies are conducted after the completion of Phase 1 studies when the safety of the product of humans is established. Phase 2 and 3 are efficacy trials. All detailed protocols describing aspects of the efficacy of the study should be submitted.

A protocol should include the following, with the specific requirements:

(a) "A statement depicting the purpose of the study.
(b) The curriculum vitae or other statement of qualifications of each investigator and the name of each sub-investigator (e.g. research fellow, resident) working under the supervision of the investigator; the name and address of the research facilities to be used; and the name and address of each reviewing Institutional Review Board.

(c) Criteria for inclusion and exclusion of the patient and an estimate of the number of patients to be studied.
(d) Design of the study and a description of methods to be used by investigators and analysts.
(e) The method for determining the dose(s) to be administered.
(f) A description of clinical procedures, laboratory tests to monitor the effects of the drug in human subjects and to minimize risk".

3.1.7 Chemistry, Manufacturing, and Control Information (CMC)

In the IND application CMC should provide sufficient information to assure qualitative and quantitative, and strength of the investigational agent. The strength of the desired dosage form shall vary from phase 1 to 2. Final release specifications for the drug substance and drug product are not expected until the end of the investigational study. The stability data is also required in addition to the CMC information.
Drug substance: A description of the drug substance pertaining to its physical, chemical, or biological characteristics is required for the planned clinical studies.
Drug product: A list of all components, including the inactive and active compounds during manufacture of the investigational agent as well as quantitative composition of the investigational drug product required.

3.1.8 Pharmacology and Toxicology Information

Information pertaining to the pharmacology and toxicology reports from in vitro animal studies is required for deciding whether to go ahead with human clinical trials, as well as the dosing that may be required. This information along with other mechanistic data is also useful as well as mandatory, in deciding the scope, duration, and the end points for a clinical investigation (USFDA 2015).

3.1.9 Toxicology

Depending on the type of the drug and duration of treatment required for a disease condition, acute, sub-acute, and chronic toxicity data needs to be generated (USFDA 2015). Toxicity profile would also include tests for mutagenicity, genotoxicity, affects on developing foetus, as well as special toxicity tests related to route of administration (e.g. inhalation, dermal, or ocular toxicology). A full tabulation of data has to be compiled in the form of dossier.

4 In Europe

In EU, biologics cover a broad spectrum of medicinal products that are biological in origin and are more complex than a synthetic one. Such a biological substance is produced by or extracted from a biological source and requires its characterization by physico-chemical-biological testing for its quality, including production process and its control. Annex II to the EU GMP guidelines are followed for the above studies (Commission E 2018). ICH S6 marketing guideline is followed like USA.

4.1 Clinical Studies

Directive 2001/20/EC, Article 9(8) of the EU, mandates that Good Clinical Practices (GCP) are to be followed in a clinical trials. The clinical trials may be conducted at a single location or at multiple locations spanning a single or more member states of the EU. The applicant has to register at the EudraCT Community Clinical Trial System and obtain a unique identification number. A covering letter quoting the EudraCT number along with a dossier containing complete documentation detailing the CMC of the product, pharmacological activities, mechanism of action, toxicological reports, clinical trial protocols, and any other information about the investigational drug, as may be required in specific cases are to be furnished to appropriate authorities in the member states of the EU, for obtaining an approval to conduct a clinical trial. Further trial needs to be permitted by the ethics committee (Council E 2001). Moreover, such an approval must not be construed as scientific advice on investigational medicinal product (IMP) (Commission E 2010).

4.2 Clinical Protocol

According to Article 2(h) of mentioned directive, the objectives, design, methodology, statistical consideration, and organization of a trial shall be decided by the sponsor's protocol and be identified by the title of the study. The protocol will refer section 6 of the community guidelines on Good Clinical Practice (ICH, ICH E6 (R1) 2002) (CPMP/ICH/135/95) containing relevant information for ethics committee's assessment.

The conditions on all the articles, i.e. Article 6(3)(a), Article 6(3)(b), Article 6(3)(g), and Article 6(3)(k) of Directive 2001/20/EC are followed in letter and spirit for designing a investigational protocol.

4.3 Investigator's Brochure

Investigator's brochure (IB) containing clinical and non-clinical data also needs to be submitted to the trial authorization agency, along with other documents as mentioned above. It also contains the clinical trial protocol, including dose, frequency/interval, methods of administration, and safety monitoring procedures covered by Article 2(g) of Directive 2001/20/EC.

The IB if requires to be updated must comply to Article 8(1), Directive 2005/28/EC, and "Requirements for authorization for the manufacturing or importation of such products and with the community guideline on Good Clinical Practice" (CPMP/ICH/135/95) prepared from all available information on clinical trials. After approval of the drug, the IB may be revised, if required, and such a document is then called "Summary of Product Characteristics" (SmPC). The SmPC should contain the safety information regarding any adverse reaction that may have been recorded in the clinical trial. If the IMP is approved only in one member state and the active substance contained in it is to be tested in a multilocational trial for approval in other countries, then the SmPC document may be used as IB in other countries where trial is to be conducted.

5 In India

In India, the regulatory authority that is designated to evaluate the safety, efficacy, and quality of drugs is the Central Drug standard Control Organization (CDSCO). Further, Review Committee on Genetic Manipulation (RCGM) of the DBT (Department of Biotechnology) has been entrusted the responsibility to oversee the development and preclinical evaluation of recombinant biologics. Recombinant biologics are regulated under the provisions of Drugs and Cosmetics Act, 1940, and the rules framed there under for the manufacture, use, import, export, and storage of hazardous microorganisms/genetically engineered organisms or cells, 1989 (Rules, 1989) notified under the Environment (Protection) Act, 1986.

5.1 Pharmaceutical Biotechnology

Along with other statutory approvals, approval on Research and Development, manufacturing import is equitably important.

5.2 Research & Development

The applicant shall submit proposal (r-DNA) to Institutional Bio-safety Committee (IBSC) with details about the organisms and vectors to be used, protocols, risks involved as well as risk mitigation strategies. If the experiments involve production in larger quantities (up to 20 l) or animal studies, then the proposal is further taken up with RCGM seeking its permission to conduct the experiment (Limited Capacity). RCGM if approves the experiments, then it is supposed to intimate its decision to State Biotechnology Coordination Committee (SBCC) and Drug Controller General of India (DCGI).

A prior approval is also required from the Institutional Animal Ethics Committee (IAEC) to conduct animal trials. Where the safety pharmacology studies are part of toxicology studies, these studies should be conducted in an accredited laboratory (GLP). These studies are to be conducted as per the OECD/ICH guidelines. After obtaining such permissions, proposal is submitted to RCGM for animal study.

5.3 Permission to Manufacture and Import of New Drugs

An application form (Form CT-12/13 (API/formulation) and Form CT-16 (Import)) shall be submitted to the CDSCO through SUGUM online portal to seek permission to manufacture new drugs for testing or analysis or clinical trials or BA/BE studies. The Central Licensing Authority may, after scrutiny of the CMC information, rationality, and documents furnished with the application, grant the permission to manufacture unapproved active pharmaceutical ingredient in Form CT-15, to manufacture its formulation in Form CT-14 and for import of new drug in Form 17 under relevant rules of Drugs and Cosmetics Act, 1940, and the permission is valid up to three years unless suspended/cancelled by the competent authority.

5.4 Clinical Trials

As mentioned in the USA and EU, detailed information compiled in IB is required in India as well, for conducting a clinical trial. Such a document contains information comprising chemical, pharmaceutical information including stability data, animal pharmacology, and toxicology data which prove the safe dose for the human subject/clinical investigations along with investigational protocol containing information about study centre, dose, frequency/interval, methods of administration, and safety monitoring procedures. The sponsor shall submit the application form (Form CT-04) (Table 1) through online portal to CDSCO, and same shall be submitted to the RCGM to seek the permission to conduct the clinical trials (Phase 1 or II or III) as per the protocol.

After successful clinical trials (Phase III), an application is submitted to the Genetic Engineering Appraisal Committee (GEAC) for environmental clearance, and thereafter, CDSCO will grant the permission to the state licensing authorities to manufacture, market/import the drug as per the Drugs and Cosmetics Act, 1940.

The summary of IND application submission in Regulatory market presented in Table 1.

6 Conclusion

It may be observed from the brief overview of the regulatory procedures presented here that in various countries around the world, including India, the main stress is of the regulatory agencies is towards safety of human subjects. The procedures undergo amendment from time to time, to keep up with the new data from R&D as well as with the technological development. The regulatory agencies also keep a constant oversight on the approved drugs available in the market. Despite the stringent regulations, in case a safety issue is not noticed during the trials, and the drug is approved and following its marketing and public use, a significant adverse reaction is noticed, it may be banned, or its use may be restricted to only certain settings. With new innovations, such as regenerative medicine, stem cell therapy, artificial organs, the regulatory procedures will continue to evolve, to allow such innovations to be used safely and in the public interest.

References

90/220/EEC, E. C., on the deliberate release into the environment of genetically modified organisms. Communities, O. J. o. t. E., Ed. European Union, 23 April 1990, vol L117, pp 15–27

Brown B (1991) Biotechnology in a global economy. US congress, office of technology Assessment, Washington, p 277

Regulations, C. o. F., 21 (2018) 21 CFR 312.22 and 312.23, investigational new drug application, services, D. O. H. A. H., Ed. USFDA, US

CFR, Instructions for Filling Out Form FDA 1571, Investigational New Drug Application (IND) (2019) Form FDA 1571, USFDA, Ed. Department of Health and Human Services, US

CFR, Part 56, Institutional Review Board (2018) 21CFR Part 56, Administration, F. a. D., Ed. USFDA, US

Commission E (2010) Detailed guidance on the request to the competent authorities for authorisation of a clinical trial on a medicinal product for human use, the notification of substantial amendments and the declaration of the end of the trial (CT-1). In: 2010/C 82/01, Commission, E., Ed. Off J European Union

Commission E (2018) Manufacture of biological active substances and medicinal products for human use, vol 4, EU guidelines for good manufacturing practice for medicinal products for human and veterinary use. Directive 2003/94/EC, Safety, D.-G. F. H. A. F., Ed. EudraLex, The Rules Governing Medicinal Products in the European Union, European Union, p 26

Council E (2001) On the approximation of the laws, regulations and administrative provisions of the Member States relating to the implementation of good clinical practice in the conduct of clinical

trials on medicinal products for human use. In: DIRECTIVE 2001/20/EC, Council, E. P. a. o. t., Ed. EU
Development, O. f. E. C. a. (1986) Recombinant DNA safety considerations. OECD, Paris
Gottlieb S (2018) M. D. a. A. A. FDA's new efforts to advance biotechnology innovation. Administration, U. S. F. a. D., Ed. U.S. Food and Drug Administration, United states
Group IEW (2011) Preclinical Safety evaluation of biotechnology-derived pharmaceuticals international conference on harmonisation of technical requirements for registration of pharmaceuticals for human use, vol ICH S6 R1, p 4
Guideline, I. H. T. (2011) Preclinical safety evaluation of biotechnology-derived pharmaceuticals use, I. C. O. H. O. T. R. F. R. O. P. F. H., Ed. vol ICH S6 (R1), pp 7, 10
ICH, ICH E6 (R1) (2002) Guideline for good clinical practice. In: CPMP/ICH/135/95, Agency, E. M., Ed. EMEA: UK, p 48
OSTP (Office of Science and Technology Policy) (1986) Coordinated framework for regulation of biotechnology. Executive Office of the President. Federal Register 51:23302. Policy, O. o. S. a. T., Ed. Executive Office of the President: NEOB-Room 5005, Washington, DC 20506
Rao SN (2018) Regulations and guidelines on biosafety of recombinant DNA research and biocontainment. In: Chapter 1, Department of Biotechnology, Ed. Ministry of Science and Technology, India, p 148
USFDA, Safety Assessment for IND Safety Reporting, Guidance for Industry (2015) 10985dft 12-14-2015, (CBER), C. f. B. E. a. R., Ed. US, p 16

Yashbir S. Bedi Professor at Academy of Scientific and Innovative Research and Scientist and Head, at CSIR-Indian Institute of Integrative Medicine, Jammu. Dr. Bedi has a long experience and strong networking skills resulting in over 100 publications and numerous patents.

Sumit G. Gandhi Senior scientist and faculty at CSIR-Indian Institute of Integrative Medicine (Jammu) and AcSIR—Academy of Scientific and Innovative Research. Working on various aspects of secondary metabolism in microorganisms and medicinal and aromatic plants since ten years. Dr. Gandhi has resourcefully collaborated with industries on various projects and is associated with many academic organizations. He did his Ph.D. at Center for Cellular and Molecular Biology (Hyderabad) focused on understanding molecular effectors and mechanisms of epigenetic regulation. His bachelors and masters were in Microbiology from Jawaharlal Nehru University, Delhi.

Inshad Ali Khan Principal Scientist at CSIR-Indian Institute of Integrative Medicine, Jammu. NASI-Reliance Platinum Jublee Award—2017 for application-oriented research in the area of biological sciences. Awarded DHR fellowship by Indian Council of Medical Research 2012 to work in NIAID, NIH, Maryland, USA. He is a member of various biological societies and academies. His areas of research have been anti-infective research with reference to identification of novel scaffolds and antimicrobial resistance and microbial efflux pumps.

Intellectual Rights

Intellectual Property Rights

Tabassum Zafar

Abstract Intelligence has no limit, and innovations have no boundaries. In this era of fast pace globalization, any information could be stolen and modified very easily. Creations of mind, whether it is an idea, thought, plan, gizmo, gadget, technology or fiction, they all covered under the legal term called intellectual property. To prevent the misuse of intellectual contents and ideas, intellectual property rights are very essential. Intellectual property rights are the legal rights, which secure the write-ups, inventions designs, methods, artistic works, innovations and intellectual ideas from being stolen by anyone else than the real owner. The registered trademark, copyright or patented idea, technology or invention remains secure for theft or being stolen. For every budding researcher and entrepreneurs, it is very important to know the scientific rights covered under the shelter of intellectual property rights. The present chapter elaborates various aspects of intellectual property rights in an interesting and concise way.

Keywords Intellectual property rights · Patent · Copyright · Scientific ethics · Trademarks · Trade secrets

1 Introduction

Intellectual property (IP) is a intangible property that is the result of human creativity and intellect. IP could be an idea, a product, a formula, a creation, a machine or service. IP covers the legal rights of any of them to the actual concept originator and protects them from being stolen or misused. Intellectual property rights (IPRs) are the rights that protect the intellectual idea or products and services based on that idea. It seems relatively novel legal branch, but it actually exists in society since long ago.

IPRs are the statuary rights that are territorial in nature for limited period protection of intellectual content for management of monopoly of commercial or noncommercial inventions and artistic works to avoid any exploitation of that intellectual idea

T. Zafar (✉)
Department of Bioscience, Barkatullah University, Bhopal, India
e-mail: tztabassumzafar@gmail.com

A. Saxena (ed.), *Biotechnology Business - Concept to Delivery*, EcoProduction,
https://doi.org/10.1007/978-3-030-36130-3_8

or invention in public domain. IPRs are the basic bottom line of commercial science as it provides legal certification of scientific ideas and concepts, which is very important in commercialization of a scientific idea or discovery. IPRs are relatively recent but an emerging topic of interest. It is still not taught under many scientific branches during the routine studies, but it is an essential requirement for researchers, businessmen and entrepreneurs in the 21st century. The present chapter is an effort to introduce and elaborate the IPRs in an interesting and understanding way (Jolly et al. 2012).

2 Historic Preview of IPR

Intellectual property rights have an interesting history in spite of their recent revival. Even in the ancient and old times, the threat for the misuse and theft of the original ideas persisted. Due to this, ancient intellectuals had developed a policy very similar to modern IPR system. In 500 B.C.E., the few popular pieces of evidence of protection of intellectual concept existed. According to this policy, only the Sybaris's Greek colony chefs retained the right to cook specific culinary cuisines. IPR in ancient times was discussed in Bruce Bugbee's formidable work. 'The Genesis of American Patent and Copyright Law,' where the incidence of Vitruvius, was discussed. Vitruvius (257–180 B.C.E.) was a poet, who recited the words, poems and verses written by any other author in a literary contest and got exposed in Alexandria. In consideration to IPR, different ownership interests were also visible among Roman (first century C.E.) painters and poets. Apart from such type of rights protection, still the ancient Greece or Rome world lacks any typical written law or institution for the convention of intellectual property protection. The intellectual ideas were however saved from being stolen by the involvement of franchises, privileges and royal favors granted. Bugbee was the person who primarily gets involved in the distinguishing between franchises or royal favors. It was distinguished that franchises and royal favors restrict the access to intellectual idea preexisting in the public domain (Bugbee 1967; Adam and Ken 2018).

The English system of American institutions of intellectual property protection initiated when the Statute of Monopolies in 1624 and the Statute of Anne in 1710 come in light. In the Statute of Monopolies, a fourteen-year monopoly was provided to the authors and inventors. However, literary works remained largely unprotected until the Gutenberg's printing press arrived during the fifteenth century. The Statute of Anne in 1710 is the foundation of the first statute of modern copyright. The law in Statute of Anne covers the protection to the author by granting fourteen-year copyrights, along with a fourteen-year renewal, if the author was alive (Farnley et al. 2004).

European countries such as Belgium, Holland, Italy and Switzerland started following the unsaid law generated by England (Bugbee 1967). The expansion of the geographic scope of intellectual property protection was spread over the globe when various international treaties like the Berne Convention treaty and the trade-related aspects of intellectual property (TRIPS 1994) come in picture.

3 Various Forms of IPR

In the 21st century, intellectual property rights can be easily classified into four major divisions named trade secret, trademark, copyright and patent.

3.1 Trade Secret

Trade secret is the confidential information related to market capitalization by which, an entrepreneur can cover the intellectual rights related to the particular secret information or industrial formula. Trade secret covers the formula process, design, practice, pattern, commercial method, instruments, marketing, export or sales strategies. A trade secret is widely used in industry and manufacturing business plans, where the rights related to the secret formula of manufactured product retained confidentially by the manufacturers using the trade secret. Trade secrets provide the economic benefit to the trade secret holders usually by commercialization (Patino 2009). One of the best example of trade secret keeping is associated with the name of a popular cold during manufacturer named "Coca Cola", This company is successfully keeping the formula as a trade secret without keeping any patent for many years. It is possible to patent someone's trade secret if found by the legitimate means. Legal circumstances related to trade secrets are less complicated in comparison with patents. A trade secret is a weak protection against the misuse, but it works with an immediate effect. Keeping the trade secret is easy and does not require any legal disclosure or registration cost but if; the trade secret is stolen, no legal action could be taken in that case. Anything that could be easily disclosed by the keen inspection of a product cannot be kept as a trade secret (Rockman 2004; Becker and Susan 2005). The secret formula of Coca-Cola is one of the best examples of trade secrets.

3.2 Trade Mark

Trademark is another category of intellectual property rights that distinguish the product, design, expression or service from a registered provider from other sources or competitors. Any phrase, logo, symbol, design, image, color, smell, or sound (like jingles) or service could be protected under the trademark policy. It is legally possible to purchase, sell or license the trademark. Trademark actually used only for the management of the goodwill of the product or service it symbolizes. There are various types of trademark that are denoted by various means. ™ is the symbol for an unregistered trademark, which is used to product or brand. ™ is an unregistered service mark, which is used to promote services provided by any particular brand. ®is the symbol for a registered trademark (Tiwari et al. 2011). All the symbols used for trademarks and service marks are written as a superscript over the brand names during commercial popularization. However, unregistered trademarks provide less protection against trademark infringement in comparison with registered trademark (Anonymous 2004).

3.3 Copyright

Copyrights are a form of IPR that protects the creative work such as books, articles, movie, short films, report, diary, songs, poems, literature, music, drama, artistic works, paintings and cinematography. Copyright could be applied through proper channel with various benefits. Copyright is mentioned by using the symbol of circled letter c (©or ©). Sound recording copyrights are mentioned by capital letter P enclosed by a circle ℗. A registered trademark used to symbolize by the capital letter R enclosed by a (®), which denotes that product is registered in the trademark office. For mask work protection of the content, a non-obligatory symbol capital letter M enclosed by a circle (Ⓜ) is used. The infringement of the copyright allows the copyright holders to receive the penalty, if the copyright was earlier registered with the original creator. The copyright is provided for a limited period of time for software. The length of copyright varies over the world. From Yemen to Jamaica, the copyright lasts minimum life plus 30 years to life plus 95 years. Copyright protection is a safe way to cover the license related to the work for the author's entire life plus seventy years after the author's death. Copyright provides various rights to the owner including, reproduction of work, distribution of work, cinematographic rights, share in economic benefits generated from the copyrighted work. However, to avoid the restricted publicity of copyrighted work in the public domain, some exceptions were considered under IPR. In order to protect the interests of users, exceptional use by non-copyright holders is considered as fair use. For educational, research, religious, criticism, performance, a limited user of the copyrighted material is considered (Alastair et al. 2010).

3.4 Patent

The patent is the form of IPRs that cover new patentable invention according to their novelty and utility under the strict monitoring and examination. According to the national law, patents' law changes for each location worldwide with some common points of consideration. A patent should be novel and useful enough for the consideration as a patent. An invention, which is frivolous or contrary to law or morality or injurious to public health, cannot be a patent. A discovery, mathematical formula, calculation or scientific theory cannot be patentable. An invention relating to atomic energy, an invention, which is in effect, is traditional knowledge cannot be patented. A method of horticulture, cultivation or agriculture cannot be patented. Apart from these, there are various such things that are not allowed to be a patent; however, some of them could be secured using other types of IPRs (Saha and Bhattacharya 2011). Patented material or process provides the inventor a kind of monopoly right gain over the invention that anyone else could not use it without asking for permission. There could be two different classes of patents termed as process patent and product patent. It could also be stated that utility patents and design patents are the subclasses available in patenting systems.

A design is defined as the 'surface ornamentation' including the configuration or shape of an object. A design patent protects only the object's appearance. It could be obtained only when the design must be inseparable from the object. A utility patent is the patent type that provides protection for the functional or structural features of the patented object. It is the most common and most useful type of patent, which could be obtained for new machines, devices, manufacturings, processes, and method. If, an inventor wishes to get patent for both the functional features and the design of the object, they should apply for utility and design patent both (Tulasi and Rao 2008).

4 IPR in Biodiversity and Bio-resources

Biological diversity includes the diverse forms of a variety of organisms, including their genetic diversity in the world. This natural biological wealth and its interrelationship of genes, species and ecosystems are precious for human society. Millions of plants, animals, fungi and other organisms are the glamorous bio-resource that we owe. Biological resources include organism's populations, genetic resources and another biotic component of ecosystems, which have significant potential use or value for humanity (Kannaiyan 1992).

It was a matter of debate that whether biodiversity and bio-resource should be covered under the terms and laws of IPR or not. The Traditional Knowledge of Biodiversity was earlier untouched by the contemporary debates related to the inclusion of these properties under the margins of intellectual property rights. The concerns about the inclusion of indigenous resources and genetic resources and bio-resources rose enormously when the use and demands for the bio-products raised with each passing year. The coverage of bio-resource under the IPR becomes a global need indeed. In the recent decades, international agreements such as the convention on biological diversity (CBD) and the trade-related aspects of intellectual property rights (TRIPS) help in the commercialization of the traditional knowledge associated with the biodiversity all over the world. Bio-rich countries, such as Third World and India, cover the countless varieties of animals, genetic resources, plants, trees, microbes which have potential applications of commercial and economic importance. To avoid the serious threat to the biodiversity and bio-prospecting, it becomes really an important aspect to cover the use of bio-resources under IPR (Utkarsh et al. 1999).

The global biodiversity pool has 1.75 million identified species among which 2.7 lakh belongs to plant kingdom. Such a huge pool of animal and plant bio-resource is not easy to handle with equal distribution and prevented misuse.

The distribution of plants and animal species is diverse and unequal across the globe. It is really very difficult to equally provide the resources among biodiversity-rich countries and areas with moderate and little biodiversity. These life forms, which are important for medicine, trade, food, fashion, tourism, energy, recreation, are nowadays covered under the boundaries of IPR. Various acts and laws under the edge of IPR in bio-resource sector cover the law related to the common heritage of mankind, biological diversity, patentable, microorganisms varieties, protection of plant varieties and farmers' rights, sovereign right of nations: conservation, sustainable utilization, equitable benefit sharing, etc.

5 IPR in the Pharmaceutical Industry

Pharma industry is a fast-growing business relative to other biotechnology and medical science-based commercial benefits. It is very competitive to maintain the quality of the drug while introducing new drugs for emerging diseases. Pharma industry has nowadays shifted its focus to long-term consumable medicines rather than acute medication. It is really costly for the company to introduce the new drug to the market, while there is a huge market full of piracy. In this fast-globalized market, it becomes essential to get a patent or IP coverage for any new or generic formula (Angell 2000; Lexchin 2005).

Over the period, the applications to cover the IPRs related to drug and pharma industry almost tripled in the last decade. This makes the competent authorities to be a little choosy for patent granting and IPR protection for new drugs. These are the reasons that drive the consequent reduction in patent protection coverage in years. The drug industry also faces many others conflicting requirements such as keeping the cost low to meet public demands, which is not an easy task because of biotechnological research and development, production, management and marketing chain (Mrudula et al. 2009). Pharma industry has a tough competition worldwide, and thus, they usually do not disclose the invention or formula by publication until they get it patented by proper channel. Patenting protects the drug formula from being stolen or copied and retains the commercial rights within the inventors and real owners. The procedure of IP protection is a bit complicated for biotechnologically developed biopharmaceuticals as compared to simple conventional drugs. Each microbial strain that is used in development should be identified before the IP protection, and if it is an existing one, then the database was considered, but if it is novel, it should be deposited and registered in the appropriate repository before patenting the drug (Glasgow 2001).

There are many limitations and hurdles in pharma industry patents as IPRs' conservation in the pharma industry is a multidimensional task. Writing patents in the pharma industry requires special expertise in professional, scientific, legal and technological expertise. The inventor could also patent new inventions, as well as new combinations, which have not used earlier. Combination of already existing substances into something useful could also be patented.

6 IPR in Bioinformatics and Computational Biology

Bioinformatics is the branch of science that deals with the computer or computer-based information technology to create the interdisciplinary and unambiguous approach to manage biological data and information. From the discovery of DNA and RNA structure, the need for bioinformatics in biological science study increased exponentially. In a modern world, the generation and management of various software-based scientific information are the biggest outcome of bioinformatics (Harrison 2003). The genesis of bioinformatics becomes more and more significant each day by development, proliferation and management of molecular science data consisting of the gene, DNA, RNA, protein and functional sequence-based data. Understanding of the biological approaches, the discovery of new drugs, detection and identification of various gene-based medicines contains lots of efforts time and funding, and thus, it becomes very essential to cover each novel and existing finding within some legal framework (Fernandez and Chaw 2005; Fernandez et al. 2013). IPR coverage of bioinformatics outcomes ensures the benefits and economic advantages should reach to the real inventors for the rest of life. The patenting system is one of the prompt tools to ensure the protection of IPRs within the bioinformatics domain. In this way, the patent containing authorities without any risk of theft, copy or misuse retains the authority to marketing and commercializing the product. Security of fundamental rights related to economic arguments is also covered by copyright protection and trademark development (Konig et al. 2015). However, there are limitations also in the case of the genome and gene-based personalized studies. Any database could be a subject matter of patenting or not depending upon the data presentation and information it consists. However, software is more likely to be a patentable intellectual property for the developers, which allows them to raise money by the commercialization. The useful, tangible and concrete information that is provided by any particular software can be covered by IPRs regime by manufacturers (McBride 2002).

Synthetic biology has grown enormously within the last few years with the development of various computational aids in research. Life constructing pathways are studied under these types of studies at the genetic level. These efforts are ahead of the scale and extent of traditional recombinant DNA technology and similar approaches. Biological engineering is a nonconventional branch that faces tremendous flaw under IPR protection. While techniques related to biotechnology have mainly difficulties for patent law, computers and computational methods both face copyright and patent problems. Copyright is considerable for original works of expression, explicitly excluding works that are functional. However, patent law considers functionality and traditionally excludes formulas and algorithms. Thus, many computerized approaches do not fulfill the considerations of neither the copyright nor the patent box. It was too functional for copyright, too close to a collection of algorithms and ideas for a patent (Gopalan 2009). Some popular foundation Biotechnological outcomes such as monoclonal antibodies and recombinant techniques either were not patented or patented. If patented, the owners of the patents were generally made the techniques and benefits available widely at a reasonable cost. But one thing is sure the law and IPRs are still struggling in between hardware and software inventions (Rai and Boyle 2007).

7 IPR in Biotechnology and Corporate Biotechnology Transfer

Biotechnology is a sister branch of biology with enormous possibilities of growth and development. Since ancient era, human beings were using applications of biotechnology in a silent manner, but within a recent decade, the development of nonconventional methods, the introduction of genetic modification in plants, animals and microbial system help this branch to bloom as a lovely flower. Because of new inventions, day by day, it is really essential for the inventors to introduce their inventions with IPRs. There are various factors that justify the essentiality of IPRs in biotechnology corporate arena.

All the common IPRs forms (trademark, trade secret, copyright, patent) are equally applicable in the biotechnology industry depending upon the subject matter. In India, IPRs in biotechnology field also face a variety of limitations. In India, it is not possible to patent any plant species. Similarly, it is not possible to patent any developed software in India, but still, the software can be secured under the functionality of copyright protection. Plant breeder's rights are the protection that covers the legal benefits of plant species, which are not possible to get patent. There are many other limitations and overcomings available in corporate biotechnology market with their existing alternative support systems (Saukshmya and Chugh 2010).

Biotechnology sector requires an updated and relevant IPRs system because of its own limitation of generating new inventions day by day. IPRs not only protect the material, invention, process, information, technology secure from unfair use but it helps significantly in revenue generation too. When a license is awarded for patented material or technology, it generates royalty for the owner in two forms. Granting permission for fair use allows the owner to receive royalty due to license sharing, but if the use is unfair, it generates compensation from the illegal users. Revenues are the biggest source of income for organizations scientists, researchers and inventors. This constant influx of funds remains in the rotation. Research and development branch requires lots of time, effort and money investment, which will recover from the patented or copyrighted product revenue. By this process, the consistent funds remain available for further survival of the industry (Giugni and Giugni 2010).

Apart from this, IPRs not only provide the financial and legal cover but it also feeds the moral and psychological reward system of the inventors. It is really worthless if the inventors have no name, fame and money associated with the hard work done earlier. Due to IPR patenting and copyright system, the individual and organization retain the rights of name, fame and revenue with them. This helps the inventors to become psychologically carefree and pleasant that they have sufficient platform to portray their skills for their own as well as social benefit simultaneously.

In biotechnology, it is possible to get patents for nucleotide sequences for a limited period of years, which then release in public domain; with a little controversy, gene patenting systems also exist in few countries. However, in general, animals, human, their genes are cells are covered by non-patentability approach. But exceptions are available. Microorganisms obtained by genetic engineering are patentable. Genetically modified crops, such as 'tryptophan over producing maize,' were patented in the USA in 1985. Similarly, the 'oncomouse' was patented in 1988, the USA, for being genetically modifiable. A plant patent is another type of nonconventional patent, which is a little different from a conventional utility patent. A plant patent is available for a plant that is obtained from asexual reproduction but not covers the tuber propagated the plant. Trade secrets are also popular IPR form in biotechnology, which is quite popular for hybridization techniques, cell line processing, customer consumer lists and corporate merchandising plans. Indian Copyright Act, 1957, has a later amendment in 1999 that reflects the Berne Convention on copyright. India is a member party of many conventions and organizations including Berne Convention, Geneva Convention, World Intellectual Property Organization (WIPO), Geneva and UNESCO.

8 Biological Database and IPR

Biological databases are the libraries like information source about existing biological data. The biological database could be classified under various sub-divisions including sequence, structural and functional database on the basis of the information. All primary (Genbank, EMBL, DDBJ, SWISS-PROT, PIR, PDB), secondary (UniGene, PRINTS, BLOCKS) and composite databases (NCBI) also have significant roles in IPR management. Primary databases are usually collected by government or social funding resources; hence, they usually open to the public without any IPR protection, while secondary databases are usually involved, capital, labor, time, etc., and thus, they are considered to be covered under IP protection (Groombridge 1992).

9 Conclusions

IPRs are very useful but only for those, who intellectually understand their right and take actions to cover it. It is very beneficial to decide early, whether any information or idea is essential to protect under the legal line of defense using IPRs. As early as the professional step taken to cover the idea, service or product, the rights become more secured in professional front from the theft or competitors encroachment. It should be chosen very carefully that what types of IPRs are useful for a particular case. Understanding the IPR is a basic prerequisite for better management and commercialization of biotechnological applications. Identification of the rights, proper planning to secure the intellectual property helps to globalize the commercialization of technology, creativity or invention without any fear of theft or misuse. IPR plays a vital role in the management of the protection of inventions and creativity. The need of the hour is to evolve IPR policies and incorporation of these into management systems as a foundation strategy. A wise decision, in the beginning, is the protector of safe and sound commercial future.

Acknowledgements Author is thankful to Prof. Vinoy K. Shrivastava, Head, Department of Biosciences, Barkatullah University and Dr. Bashirulla Shaik, Assistant Professor, National Institute of Technical Teachers Training, Bhopal, for their valuable guidance, consistent support and blessings.

Conflict of Interests Author declares that there is no conflict of interest.

References

Adam M, Ken H Intellectual property. The Stanford Encyclopedia of philosophy (Winter 2018 Edition), Zalta EN (ed). https://plato.stanford.edu/archives/win2018/entries/intellectual-property/. Available at https://plato.stanford.edu/entries/intellectual-property/

Alastair MJ, Martin K, Lionel B (2010) A mongrel of early modern copyright. In: Deazley R (ed) Privilege and property: essays on the history of copyright. Open book publishers. ISBN 978-1-906924-18-8

Angell M (2000) The pharmaceutical industry. To whom is it accountable? N Engl J Med 342:1902–1904

Becker SJ (2005) Trade secrets and confidential commercial information. Discovery from current and former employees. American bar association. ISBN 978-1-59031-485-2

Bugbee B (1967) Genesis of American patent and copyright law. Public Affairs Press, Washington, DC

Farnley S, Morey-Nase P, Sternfeld D (2004) Biotechnology—a challenge to the patent system. Curr Opin Biotechnol 15(3):254–257

Fernandez D, Chow M (2005) Intellectual property strategy in bioinformatics and biochips. Biosens Bioelectron 15, 21(1):197–200. https://doi.org/10.1016/j.bios.2004.09.014

Fernandez D, Maninang A, Kobayashi S (2013) Intellectual property protection for bioinformatics and computational intelligence. In: IEEE symposium on computational intelligence in bioinformatics and computational biology (CIBCB). Singapore,pp 31–35. https://doi.org/10.1109/cibcb.2013.6595385

Giugni D, Giugni V (2010) Intellectual property: a powerful tool to develop biotech research. Microb Biotechnol 3(5):493–506. https://doi.org/10.1111/j.1751-7915.2010.00172.x

Glasgow LJ (2001) Stretching the limits of intellectual property rights: has the pharmaceutical industry gone too far? IDEA J Law Technol 41:227–258

Gopalan R (2009) Bioinformatics: scope of the intellectual property protection. J Intell Prop Right 14:46–51

Groombridge B (ed) (1992) Intellectual property rights for biotechnology. In: Global biodiversity: status of the earth's living resources. Chapman and Hall, London, pp 495–499

Harrison R (2003) Protecting innovation in bioinformatics and in-silico biology. Bio Drugs 17(4):227–231. https://doi.org/10.2165/00063030-200317040-00001

Jolly M, Fletcher AC, Bourne PE (2012) Ten simple rules to protect your intellectual property. PLoS Comput Biol 8(11):e1002766. https://doi.org/10.1371/journal.pcbi.1002766

Kannaiyan S (1992) Potentiality of Azolla as biofertilizer in sustainable rice farming. In: Swaminathan MS, Jena S (ed) Biodiversity and implications for global food security. Macmillan Publications, Madras, India, p 326

Konig H, Dorado-Morales P, Porcar M (2015) Responsibility and intellectual property in synthetic biology: a proposal for using responsible research and Innovation as a basic framework for intellectual property decisions in synthetic biology. EMBO Rep 16(9):1055–1059. https://doi.org/10.15252/embr.201541048

Lexchin J (2005) Intellectual property rights and the Canadian pharmaceutical marketplace: where do we go from here? Int J Health Serv 35:237–256

McBride MS (2002) Bioinformatics and intellectual property rights protection. Berkley Technol Law J 17:1331–1363

Mrudula BS, Durgadevi NK, Madhavi BR, Tejeswi B, Durga PV (2009) Intellectual property rights pinpoint at IPR spotlights coveted R and D. Drug Inv Today 2:197–201

New Delhi: Commercial Law Publisher (India) Pvt. Ltd. (2004) Anonymous. The Trademarks Act 1999 along with trade Marks Rules 2002

Patino R (2009) Intellectual property rights and research disclosure in the university environment: preserving the commercialization option and optimizing market interest. JAALAS 48(2):138–143

Rai A, Boyle J (2007) Synthetic biology: caught between property rights, the public domain, and the commons. PLoS Biol 5(3):e58. https://doi.org/10.1371/journal.pbio.0050058

Rockman HB (2004) Nature of a trade secret. Intellectual property law for engineers and scientists. Wiley-IEEE. 25.3. ISBN 978-0-471-44998-0

Saha CN, Bhattacharya S (2011) Intellectual property rights: an overview and implications in pharmaceutical industry. JAPTR 2(2):88–93. https://doi.org/10.4103/2231-4040.82952

Saukshmya T, Chugh A (2010) Intellectual property rights in synthetic biology: an anti-thesis to open access to research? Syst Synth Biol 4(4):241–245. https://doi.org/10.1007/s11693-011-9067-6

Tiwari R, Tiwari G, Rai AK, Srivastawa B (2011) Management of intellectual property rights in India: an updated review. J Natural Sci Biol Med 2(1):2–12. https://doi.org/10.4103/0976-9668.82307

TRIPS (1994) https://www.wto.org/english/tratop_e/trips_e/intel2_e.htm.

Tulasi GK, Rao BS (2008) A detailed study of patent system for protection of inventions. Indian J Pharm Sci 70(5):547–554. https://doi.org/10.4103/0250-474X.45390

Utkarsh G, Gadgil M, Rao PRS (1999) Intellectual property rights on biological resources: benefiting from biodiversity and people's knowledge. Curr Sci 77(1):1418–1425

Dr. Tabassum Zafar is a guest faculty and researcher currently associated with Department of Biosciences, Barkatullah University Bhopal, Madhya Pradesh, India. She is the recipient of various fellowships including CSIR research internship award by Council of Scientific and Industrial Research, India, DST-Inspire fellowship by Department of Science and Technology, Government of India, FTYS—fellowship for training of young scientist by Madhya Pradesh Council of Science and Technology. She has also received many Young Scientist Awards and felicitation from Madhya Pradesh Council of Science and Technology, Royal chemical society of London, (North India Chapter) and society of Lifesciences.

Commerce and Management

Evolution of Biotechnology as a Million Dollar Market: The Management and Commerce of a Biotech Start-up

Gaurav Verma and Srividhya Ravichandran

Abstract This chapter focuses on the commercial and management aspect of biotechnology's business in today's scenario. The recent trends and the developments in the business models of the biotech sector would be dealt in detail. The recent economy predicts the pharmaceutical sector as the most capital intensive market area but the challenges in taking up a career in biotechnology and ethical principles that govern the setting up of an industry in core areas of biotechnology are huge. The role of biotech industry in terms of commercial, industrial and economical grounds gains importance since it requires a lot of manpower in the intellectual and in the commercial front. A success story of a biotech sector involves a lot of challenges to be faced. Hence, the chapter would highlight the challenges that equip oneself to be an entrepreneur amidst the growing demands of the biotech sector. The chapter also throws light on the core and sub-domains of life science and biotech companies which have grown enormously in the past two decades.

Keywords Biotechnology · Health care · Industrial biotechnology · Gene editing tools · Plant breeding techniques · Bioprocess engineering · Entrepreneurship

1 Current Perspectives in the Biotech Sector

Biotechnology is recognized as one of the emerging areas in science and technology having significant potential with a wide array of applications in health care, agriculture, process industry and service sectors. Biotechnology is the future of the country, and the private sector is the driving force of biotechnology supporting industrial R&D and production capabilities which in turn greatly influence the human resource development (Ninawe 2019). The ultimate aim of biotechnology is to deliver materials, molecules or processes of interest for the medical, agricultural and environmental

G. Verma (✉)
Clinical Research Center, Lunds University, Malmo, Sweden
e-mail: geneverma@gmail.com

S. Ravichandran
Department of Biotechnology, Indian Institute of Technology, Chennai, India

A. Saxena (ed.), *Biotechnology Business - Concept to Delivery*, EcoProduction,
https://doi.org/10.1007/978-3-030-36130-3_9

market. The past role of a biotech company has been more or less like an intermediary between science and commercial applications (Joshi 2018).

The biotech industry, just like any other industry, is dedicated towards product development but the difference is that there are stringent rules monitoring the production (Kaitin 2010; Khilji et al. 2006; Maak and Wylie 2016). The amount of money invested in the industry is very high while an uncertainty always remains about the end profits (Chok and Sun 2007). With passing time, this industry continued growing by academic collaborations and in-licensing of valuable research outcomes (Stuart et al. 2007) accounting for the biotech industry which seems to be an important part of modern economy. There have been intensive investments in this sector globally, especially in 2015 where this sector received the second highest funding and hence rapid generation of business models (Segers 2015). Learning is important in industries such as biotechnology in which knowledge is developing rapidly and the focus is on new innovations and is indeed often a primary motive for entering into partnerships.

2 Fields of Biotechnology

The different fields of biotechnology can be classified by colours, as a "rainbow" methodology (Kafarski 2012) (Fig. 1)

In this sense, the several areas of biotechnology identified and classsified include:

- white—industrial biotechnology
- violet—law, ethical and philosophic issues
- blue—marine (aquatic)
- green—agriculture
- yellow—nutritional biotechnology
- red—medicine and human health
- grey—environmental protection
- brown—desert and dry regions
- gold—bioinformatics, computer science and chip technology
- dark biotechnology—bioterrorism and biological weapons.

There are some areas like nanotechnology which are grooming vigorously which are not coded by the above classification; yet, they stand out vibrantly in the arena of biotechnology. The industry could be broadly divided into pharmaceutical and non-pharmaceutical sectors (Horvath et al. 2019). The pharmaceutical companies have been on the top list in terms of the investments made since the evolution of a new drug takes its time and money. Hence, the non-pharmaceutical companies find a better position in the industry itself.

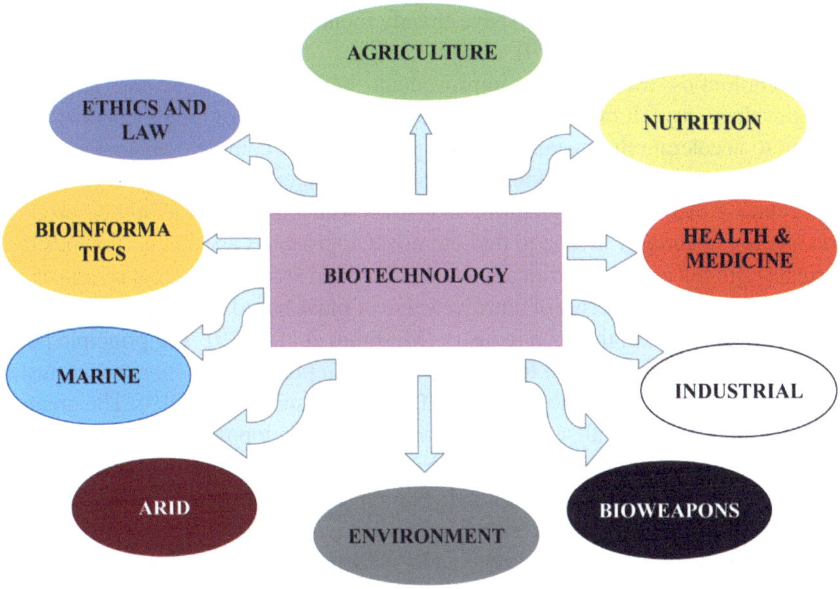

Fig. 1 Areas of biotechnology. The figure represents various streams of biotechnology as given by their color codes; green-agriculture; yellow-nutrition; red-health and medicine; white—industrial; purple—ethics and law; gold—bioinformatics; blue—marine; grey—environment; brown—arid and black—bioweapons

3 Commerce and Management of the Agricultural Biotechnology

Genetic engineering has revolutionized food and agriculture to an enormous level. Gene manipulation techniques have rendered many useful products like nutritional supplements, food flavouring substances and food additives which are useful to enhance the standard of the food industry to greater extents. Metabolic engineering plays a progressive role in the production of different types of genetically modified foods in the food industry. The development of functional foods plays an important role in the area of food biotechnology. The presence of secondary metabolites and also the genetic engineering of the microorganisms to incur the production of useful products have been under research for the past three decades. The development of metabolomics and metabolic engineering is some niche areas where the green revolution has really paid off in the recent past.

The field of plant biotechnology is being transformed by *new plant breeding techniques*, which is an orchestration of novel genetic techniques. There are tools available to manipulate DNA at targeted sites to achieve specific traits. Of these tools, clustered regularly interspaced short palindromic repeats (CRISPR) —CRISPR-associated protein (Cas9), transcription activator-like effector nucleases (TALENs) and zinc-finger nucleases (ZFNs) have already found widespread use in research and development (Kleter et al. 2019).

Metabolomics, used for crop improvement, is one of the most emerging technologies these days. It facilitates study of complex networks of biological processes, aiming to accelerate the sustainable crop production. A complete understanding of the enzymes present in traditional food and their nature of presence is extremely important for the food manufacturing industries which are involved in their production and application. It is time to find out novel approaches and easier methods that mitigate with the genetic modification techniques and are also expected to contribute to the value addition in terms of nutritive value of plant-based food and also the quality of enzymes used in the food industry (Meshram et al. 2019). The principle point of GM crops is to accomplish food security or to improve nutrition, enhance sustenance and advance manageable farming (Chaudhary and Singh 2019). The greatest benefit of the agricultural sector is the technology development and genetic engineering techniques to bring in herbicide-tolerant and sustainable crops. However, at times, the social and ethical dilemmas drive the industry very hard but GM crops are already into human consumption after successful field trials.

The products produced by the genetically modified microorganisms are in use for the promotion of health benefits and prevention of malnutrition problems. For example, it is clear from a report that there are about 3000 different species of microalgae which have the potential to produce triacylglycerol which are known precursors for biofuels. Hence, focus would be laid on tapping the potential of biofuel generating microalgae to capture the current market. Inventions are in progress to generate and modify better molecular tools and technologies to instigate the process of synthesizing triglycerides in lipid-accumulating microalgae. The quality and quantity of agricultural products obtained using biotechnology are huge. The products in commercial biotechnology are also on the rise every year. In other words, biotechnology has yielded unique solutions for sustainable environment-friendly products through industrial development. As a benchmark, biotechnological innovations must owe to provide sustainable solutions for environmental, economical and cultural problems apart from providing nutritive food. Figure 2 illustrates the happenings in the agricultural and nutritional biotechnology.

Continuous demand for agricultural growth in developing countries and the lack of supply owing to less productivity have affected agricultural market trends in a large scale. The contributory factors for loss of balance in supply-demand cycle are the dwindling natural resources and the change in the global climatic conditions. Small-scale farming had been the highest per cent of agricultural suppliers in developing countries where agriculture becomes the major occupation till date. Agricultural biotechnology has really contributed by contributing new ideas, techniques and processes to keep up their roles and also by yielding sustainable agriculture solutions to small farmers. Nevertheless, GM crops have been proven to be disease and herbicide resistant and also earned a label for global identification for the economic growth of the agricultural biotechnology sector. Science and technology remains the sole problem solver amidst adverse climatic, social, economical and environmental problems being imposed on the farmers. Research and development at the field level along with

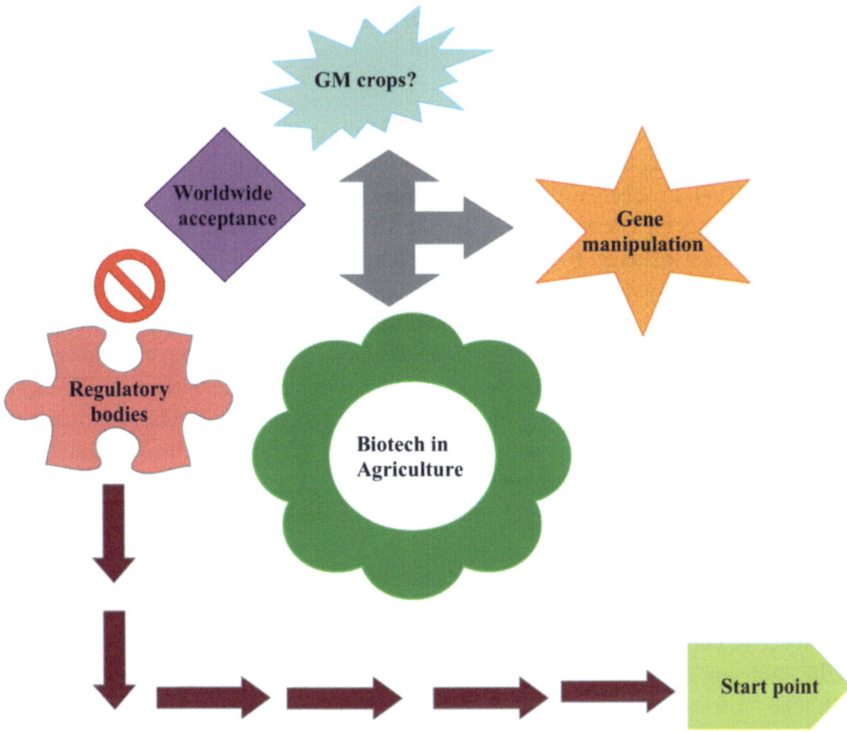

Fig. 2 Status of biotechnology in agriculture. Biotechnology in agriculture encompasses gene editing and manipulating tools which results in the rise of the GM crops but the success of the same is not likely to address its problems of worldwide acceptance owing to the regulatory compliance of the industries and processes involved

information practices should reach the farmers for them to get hands-on training and solutions to develop their capacity. Public governance and innovation governance play an important role in the commercialization and globalization of the GM products. There must be some decisive changes at the policy and operational level to bring out new financial policies which strengthens public–private partnerships that could initiate a global cascade of changes and a positive transformation of research and development, product development and technology transfer from small- to medium-scale farmers which would improve the economy of the agricultural biotechnology sector and render a greener solution for the existing hurdles which might turn out to be big barriers of growth of the sector in the long run.

4 Commerce and Management of the Industrial Biotechnology Sector

Industrial biotechnology involves the use of microorganisms to synthesize a number of products impacting our lives and life spans and is valuable to human beings (Vijay-Paul et al. 2019). Products used in daily routine are the major focus of the industrial biotechnology sector (Pessoa et al. 2019). FMCG industry has its development on a daily basis. For instance, food, cosmetic, pharma manufacturing and related fields are looking for alternative approaches to sustain their market in a rather sustainable way. The supply-demand cycle is the main criterion which decides the market trend. We must also realize the fact that there is an increasing demand from consumer market for bio-degradable and natural products instead of synthetic derivatives. Today's industrial biotechnology is ready to face the real challenges imposed by the consumer end and also enter into the business in a full swing to meddle with the trend and also take care of the supply-demand cycle.

Recent advancements in the genetic engineering techniques have paved way for commercial production of food products which are consumed widely across the world. Promoter engineering is an example of such technique to regulate gene expression and optimize metabolite biosynthesis in metabolic engineering and synthetic biology. The gene expression techniques which involve direct or cascading effects on the multi-gene pathways could significantly contribute to the metabolic flux distribution and also to the maximization of the production of certain metabolites which are real need of the hour (Xu et al. 2019). The end product of gene modification could be of direct utilization in food, pharma and other industries which retain the consumer market in an equipped manner.

Metabolic engineering, enzyme engineering, genomics, proteomics, metabolomics, etc., play an important role in the process of intriguing the genetic potential of the microorganisms to increase the yield of useful products like secondary metabolites, nutraceuticals and essential phytonutrients, biopolymers, pharmaceuticals and clinically useful enzymes. Several advancements in the industrial biotechnology corridor have opened up new vistas for budding biotechnologists who would get an opportunity to work under any of those branches. Right from the internship level, opportunities have been huge for the future biotechnologists who aspire to take up any research and developmental activities in any of the industrial areas related to biotechnology. Figure 3 clearly depicts the prospects of the industrial biotechnology with reference to the industrial demand of the novel products in the market.

Industrial biotechnology has a lot of scope, and its potential contribution towards economic growth is very huge and is supposed to pave way for a sustainable global economy in the future. As discussed in the previous segment, academia-industry partnership can strengthen the industrial biotech sector in a very good manner. Consultancy, contract research, public–private partnerships and bilateral partnership firms are at their good deals to open up their economy stride into the industrial sector. In the industrial biotech sector, few small and medium sized enterprises (SME) do exist

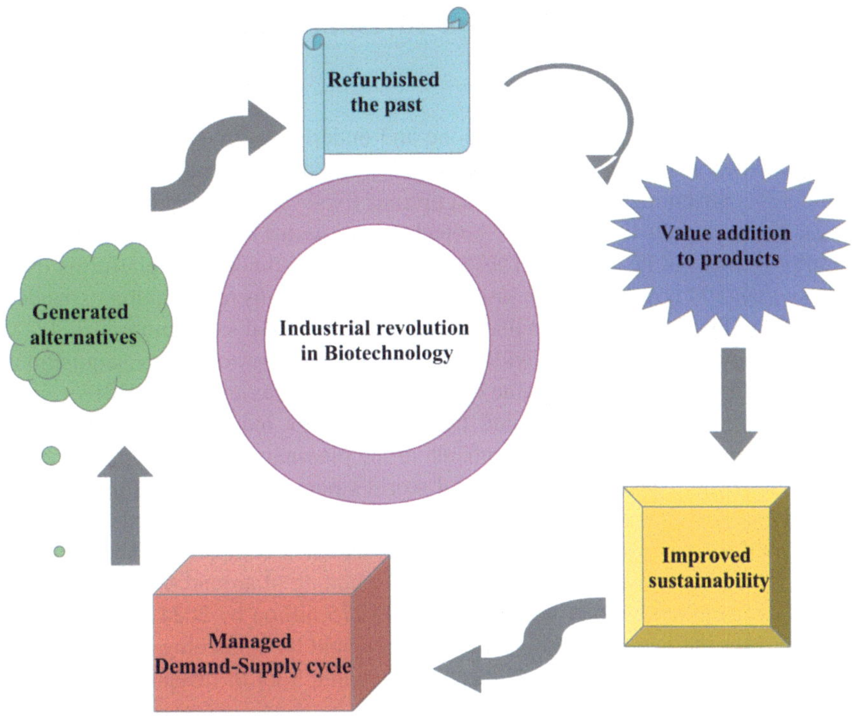

Fig. 3 Industrial biotechnology—the past, present and the future. Industrial biotechnology has always been an indispensable tool in meeting the needs of the consumers. The products rendered by industrial biotechnology have added value, improved sustainability and kept the supply-demand cycle at a threshold

when compared to the biopharmaceutical sector. These SMEs are in partnerships with multinational companies. Biotech start-ups are a welcome note to the industrial biotechnologists where the academia/universities are in a collaborative mode of working. The technical know-how is transferable, and the ideas could be incubated into these start-ups which would then develop into an independent firm. To create an ideal business model is very difficult with the start-ups but the academia-industry partnerships have definitely improved the success of the industrial biotechnology sector to the next grade. A huge cost of commercialization and a lack of marketing knowledge are barriers to technical work up involved in the technology transfer process happening with biotech start-ups. A completely niche product would definitely cross all the barriers and be successful in the industry. Hence, it is mandatory that the start-ups must think in the ways of the consumer as well as the producer of the same.

5 Healthcare Industry a Major Investment Corridor of Biotechnology in Future

Innovation has always been the backbone and underlying strength of the pharmaceutical industry. During decades, the industry has delivered multiple life-saving medicines contributing to new treatment options for several medical needs (Khanna 2012). The accomplishments of the biotechnology sector have been formidable in the improvement of health care. These include 16 biotechnological platforms used in more than 200 novel products, more than 250 indications representing expanded and improved patient care, more than $160 billion in global sales, more than 4,000 biotechnology companies worldwide and 170 biotechnology company acquisitions by pharma firms represented a value of $185 billion (Evens and Kaitin 2018).

The pharmaceutical sector bound by the healthcare industry has grown enormously in recent years. The developments in the pharmaceutical sector are vividly seen by the drug expenditure of the total world's population which is on the high rise and expected to touch 1.5 trillion $ by the end of 2021, and also the revenues gained by the sector is touching their peak levels when compared to the previous decade. There is a prospect of growth in drug spending of around 6% by 2021, the global biotechnology market will surpass USD 775 billion by 2024, and the areas earmarked for the increased drug expenditures would be in oncology, diabetes and autoimmune disorders. In order to meet the demand, a lot of inventions would be expected in these areas in the future also.

The major forecasts for the healthcare sector are on the rise since the healthcare costs for chronic ailments like cancer and neurodegenerative disorders are already peaking. The demand for effective drugs and vaccines is the need of the hour. These trends would be accommodating the drug market which is driven by pharmaceutical biotechnology sector. Currently, nearly two-thirds of the funding for biotech research and development comes from the pharmaceutical industry, while over one-quarter comes from government sources, and less than four per cent comes from universities (National Science Foundation).

The global recognition of the pharmaceutical industry is incomparable, and it has reached its peak in the past decade and continues to be a powerful industry all the more. Nevertheless, the European and American countries remain to be the major producer of products formed from the entrepreneurial biotechnology sector. This includes biotech-based products in health care whose value would be around 70 billion US dollars by 2020 which is a huge sum and reveals a dependency of the global market on these service providers like US, UK and Germany. To mention, a noteworthy contribution towards health care is explicit when the utilization of recombinant proteins expressed in plants has been a novel approach in biopharmaceutical industry that renders practical as well as safety advantages over traditional approaches (Moustafa et al. 2016). During decades, there have been several players (4300) in pharma sector but only a small subset (261) seems to have tasted success with at least one NME approved (Munos 2009).

6 Commerce and Management of the Pharmaceutical Biotechnology Sector

Healthcare industry is facing major challenges owing to the dynamicity of the market trends in this sector. It has become a capital intensive market, and major hurdles faced by the pharma industry are huge cost of investment towards research and development right from the start-up stage to fully developed MNCs. Even in developed countries, venture capitals face a setback owing to the high investment barrier in this industry. In order to keep up with the market trends and stay profitable, companies are trying to build up their economic scales. The scenario is really worsening in a long timescale. Much attention is needed to keep their heads up in the gearing situation and to take bold initiatives to be in the game. Drug discovery costs are huge, and big pharma companies are not able to cope up with the investment challenges of the same.

Neurodegenerative diseases and rare genetic disorders are imposing challenges to the pharmaceutical leaders to invest huge sum of money towards R&D in the drug discovery market, and the economy seems to be much profitable when compared to the past. Amidst the problem of expiring patents, pharma leaders are trying to strike a balance between the achieved profits and new investments by carrying out basic research and focus on generics diversification and collaborative research. The drug development costs are at their skyrocketing levels in the past two decades. In order to comply with the rules and regulations of a pharma manufacturing practices, increased cost of production, scale-up and regulatory approval processes of new drugs, pharma companies are facing a lot of challenges. Many large biopharma corporations are trying to operate in unison with the local biopharmaceutical industry situated at their offsites in any developing economic countries to perform manufacturing in these countries, with local contract manufacturing organizations (CMOs) and trying to be a solution provider for huge investment and long duration process of biopharmaceuticals. Merger of small pharma and medical biotech companies into one and acquisition of pharma companies by profited multinationals are small stop-gap solutions foreseen.

7 Nano Biotechnology, Promising Era of Biotechnology

Nanotechnology is established on the design, development and application of materials and devices possessing at least one dimension sized in a nanometre scale (Assa et al. 2016). Nanotechnology became one of the leading sectors of biotechnology that contributes to the global rise in the market and has risen to become a trillion dollar industry in the recent past. The predominant areas of use of nanotechnology in medicine, classified as nanomedicine, are in drug delivery and as diagnostic imaging and biosensors. Nanotechnology is also emerging as an indispensable tool in various parts of the biotechnology industries. The nanomaterials which are developed using nanotechnology could be used instead of chemical pesticides, fertilizers

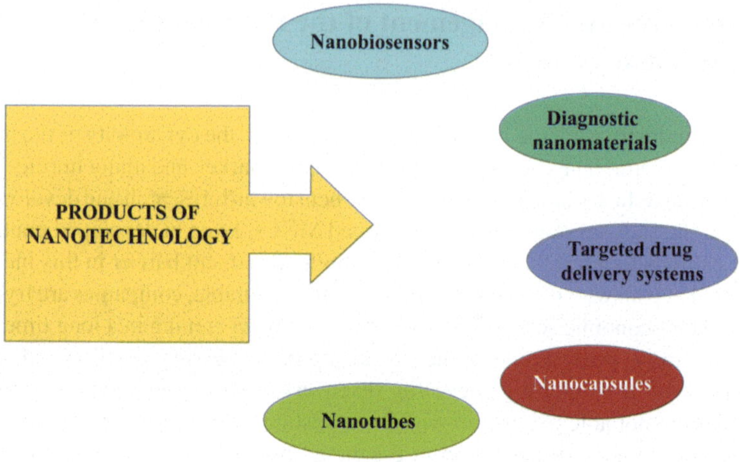

Fig. 4 Products of nanotechnology. Nanotechnology has grown as the biggest field of biotechnology and the products given by nanotechnology include nanoparticle based biosensors, nanotubes, diagnostic nanomaterials, nanocapsules and targeted drug delivery systems

and herbicides by promoting green and sustainable agriculture through the use of nanofertilizers, nano pesticides and detection and control of plant diseases by using nanoparticles; development of diagnostic tools for detection and control of human diseases (Rohela et al. 2019). Figure 4 represents the products that are obtained from the development in nanotechnology.

These diagnostic tools could have implications in molecular diagnostics, drug discovery processes, targeted drug delivery, lab-on-chip, tissue engineering and regeneration. High-throughput microfluidics has revolutionized biotechnology assays, enabling intriguing new approaches often at the single-cell level (Nagendran et al. 2018). Targeted delivery of drug using nanoparticles has always been a theme of interest in nanoscience and nanotechnology. Scientists have been trying hard to use these nanodelivery systems to target cancer cells and tumour tissues which otherwise would proliferate to a greater extent even with conventional protocols for treatment. A complete utilization of the nanoparticle- and nanomaterial-based approaches would enhance the future technology and determine the fate of the market in agriculture. Production of products based on nanotechnology has also faced the hurdles in their initial phases, and currently, there is a welcome note added to the inventions from the nanotechnology corridor. Futuristic innovations might include nanorobots for performing surgery or as artificial organ substitutes and a lot more inventions to come.

8 Career Paths for Biotechnology Students

Biotechnology has proved to be a viable vehicle for the development and utilization of technologies, which has brought not only advances to society, but also career opportunities to nation-states that have enabling conditions (Dettenhofer et al. 2018). Because biotechnology has applications in many industries, professionals can choose to work for a variety of organizations, including government agencies, private companies, regulatory bodies or clinical laboratories. Biotechnology employers range in size and type from small start-ups to global pharmaceutical leaders. Employed by medical communications agencies, contract research organizations and pharmaceutical companies, medical writers distil and translate complex clinical and scientific data to develop documentation spanning the entire pharmaceutical product life cycle, from clinical development to registration and marketing (Hager 2019).

Though the biotechnology industry is a major economic driver, generating approximately \$140 billion in revenue and also the demand for skilled professionals will continue to rise, to open up a career immediately after biotechnology in the core field is yet another challenging task for biotech/life science students. As the industry is still not self-sufficient, the students must be more dynamic and open towards the opportunities which are coming to them soon after their degree. A postgraduate diploma course in applied fields/information technology would help them mend their candidature towards a good opportunity in biotechnology. Figure 5 shows the various career paths of a biotech student in a pictorial form.

Flexibility and adaptability are two important mantras that every biotech student has to keep in mind. If they choose to work in the core area of biotechnology, there will be a limited number of options available. Henceforth, the students must be smart enough to make themselves strong to choose their career option in biotechnology and its allied fields.

9 Entrepreneurs in Biotechnology—Current Scenario and Future Prospects

According to many experts, over the next few years, an innovative-driven or knowledge-based economy, in which the production and the use of knowledge and innovation as a source of gaining wealth and competitive advantage, will replace the current traditional economics (Tohidyan Far and Rezaei-Moghaddam 2019). Researchers have shown that entrepreneurship-related education could not leverage the process of transformation of additional products into growth, and founders who start their businesses immediately after completing their studies might struggle when they seek growth by adding more products.

The entrepreneurial experience would be adopted from their co-workers or a known guide who could instigate the entrepreneurial ways to them. The ecosystem for successful start-ups typically starts with proper mentorship with a mix of

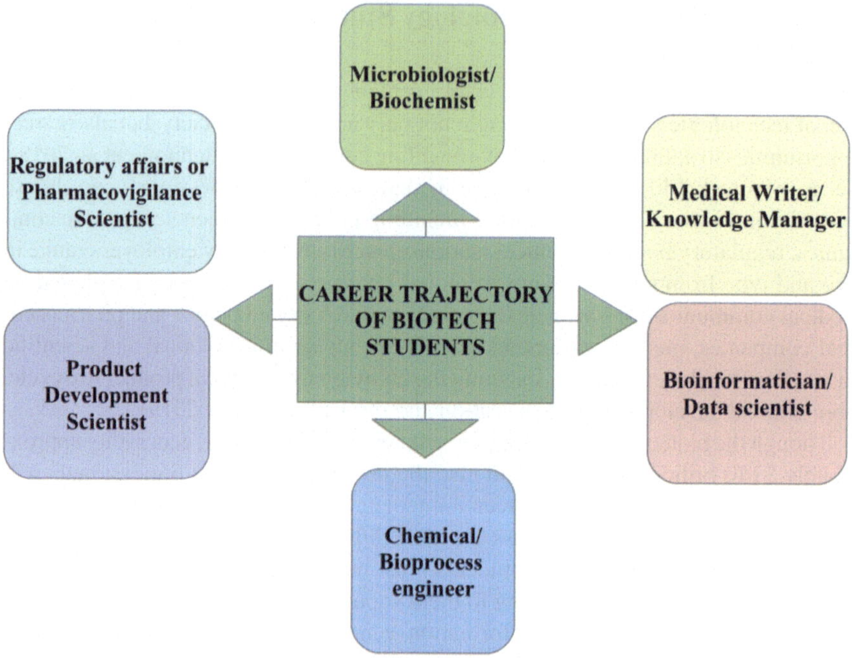

Fig. 5 Career prospects of a biotech student. A biotechnology student can acquire several career prospects depending upon his/her area of expertise. Some of the jobs include microbiologist, medical writer/knowledge management expert, pharmacovigilance/regulatory affairs expert, product development scientist, a bioprocess engineer or a data scientist

technological and directorial support focusing on the career trajectory of biotechnology (Munshi 2019). Scientists from US universities find that the number of job changes is positively related to the likelihood to become an entrepreneur, as individuals with many job experiences have had frequent employment offers and are therefore assumed to possess valuable skills (Stuart and Ding 2006).

The hallucinations and the imaginative thought processes about the pros and cons of the biotechnology industry obtained in the past would be deceiving to the forthcoming and budding entrepreneurs, as the industry itself is dynamic and unpredictable. The commercial part of biotechnology lies in the core domains of the subject itself while entrepreneurship and business environment differs across countries (Cardon and Kirk 2015) and contextual factors such as cultural, social and economic environments highly influence entrepreneurship research (Zahra 2007).

Some small towns and semi-cities have started to witness a revolution in the growth of entrepreneurship, and there are people who are taking very traditional businesses with a new dimension. One such example is biotech entrepreneur Sarah D'Souza, Founder and Chairman of Biosyl Technologies, a biotech-based company, headquartered in Hubballi, which is a small town in India which deals with both manufacturing and services. Entrepreneurial activity involves a lot of newer technologies flowing in, which are often found as complex process formed by the co-evolution of technology commercialization and entrepreneurship (Giones and Brem 2017). A smaller technology might get a warmer note when it begins its real commercialization into the market. There are times when a product developed by a huge venture capital has failed to survive in this big arena, owing to the technology commercialization and increased changes in the market. Hence, a vibrant situation and a trendsetting attitude towards the entrepreneurial approach are the deepest truth of the market that every entrepreneur has to keep in his mind. For an easy understanding, readers could refer to Fig. 6.

This era is marked as a knowledge era as the intellectual capital plays an indispensable role in the economy and growth of any market in a global pattern. Surplus economic development and technological advances are seen in the areas such as pharmaceuticals, aerospace and telecommunications. Though these areas are widely separated in their distribution and consumerism, knowledge becomes a key factor in controlling their growth levels (Lastres and Sarita 1999). Without an intellectual capital driven by highly motivated entrepreneurial skills, a start-up would not exist in the present economic situation.

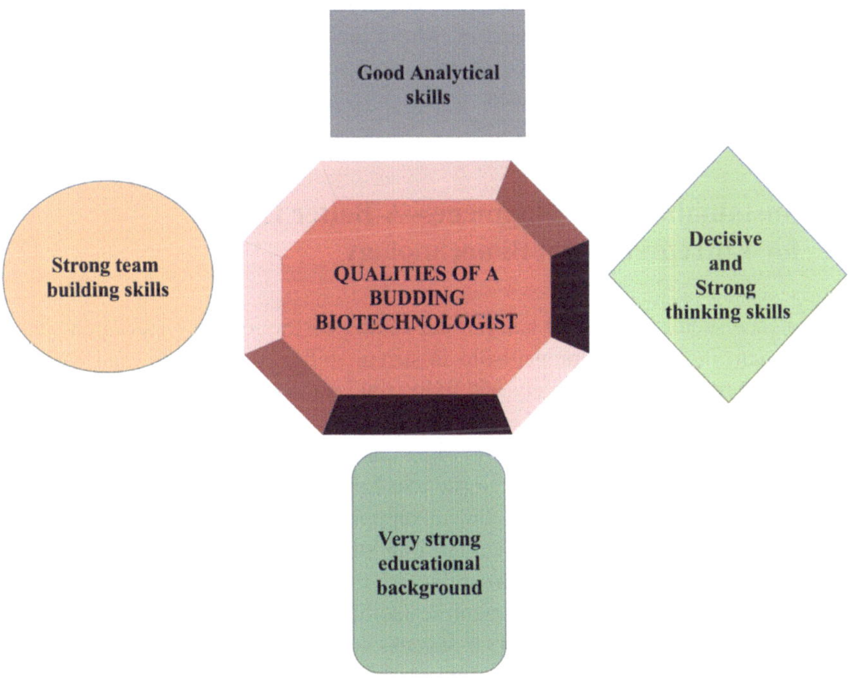

Fig. 6 Qualities of a budding biotechnologist. A biotechnologist who wishes to become an entrepreneur has to imbibe certain qualities like a good thinker and a decision maker, an expert in analytical skills and a strong team player

Support from educational and research institutions would also lead to the birth of an entrepreneur from any moderate student who wants to survive in the core industry. The shortcomings in entrepreneurial development are addressed to the issues owing to the lack of conceptualism, understanding of how a few core institutions influence entrepreneurship and the negligence of the scholars towards incrementally innovative ventures (Dilli et al. 2018) which constitute a distinct (and under-researched) type of entrepreneurship next to the (over-researched) form of radically innovative, high-growth or high-tech entrepreneurship.

In order to find a business environment and capital friendly, a funding source becomes indispensable. We could see that the Indian industry has faced a lot of challenges in terms of funding. The government has also been a crucial factor in such states where there is a real lack of other sources like venture capitalists who face the real challenge of the market risks all the time. Today's complex problems require multidisciplinary approaches and team science, with investigators who are equipped with sophisticated data analysis skills, expertise in ethical research and other advanced capabilities (Katz and Glass 2019).

Biomedical research faces a scarcity of scientists who are able to work effectively across diverse scientific disciplines (Lamb and Curtin 2019). The opportunities that develop in these research areas are huge which need a strong interaction between the academic and industrial researches, where each side is open to collaborate and learn from the other and also focuses on those research fields where the individual partners can provide real excellence.

10 Sustainability Development—A Better Option for Entrepreneurial Biotechnology

Life science sector shows rise and fall depending on the industry. Certain areas of development include entrepreneurship in sustainability development and environmental management. The society and industrial sectors are facing important challenges regarding the production of bioproducts influenced by social responsibility and environmental consequences (González-García et al. 2019). It would be nice to recall the story of a life sciences entrepreneur Edward Robinson and his colleagues who were seeking financial and technical support to develop a radical approach to fish farming. The main aim of their farming strategy was to create a zero-waste sustainable facility that would be friendly with the ecosystem. Their magnum opus fetched them the state-of-the-art scientific facilities and also some guidance from faculty experts and researchers. Their success story continued to secure $660,000 from a national funding agency that supported more growth.

One of the primary goals of bioeconomy is the development of new bioproduction systems to produce fuels, chemicals and other materials at large scale from renewable resources or various feedstocks including industrial waste streams using bio-based conversion technologies (Zeng 2019). Converting generated biomass (which has the capacity to grow on wastewaters) from microalgae into the product such as pharmaceuticals, biofuels and food supplements could also bring in greener solutions for the major challenges faced today. Also these products would be sustainable and environment-friendly and also cope up with the increasing demand of the food and energy needs of mankind.

There is an increasing demand of oil products, and substitution of the same with bio-based products would pave way for the emergence of a new bioeconomy and new industrial processes paying tributes to the sustainability development concept. Industrial biorefinery can be developed on the principle that any residues of one can then be exploited as raw material by others in an industrial metabolism (Octave and Thomas 2009). Post-bioremediation yields the chance of marketing high value-added products obtained from the agricultural waste. Such type of sustainability ventures would face a welcome note by the people per se. Process biotechnology industries which are involved in bioremediation could also offer employment to self-help groups and create an opportunity to the people around the place. Some start-up ventures involved in the management of agricultural waste using microalgal sources stand as a good example of this.

Methods of genetic engineering which include genome editing or gene transfer would easily overcome the adverse environmental conditions by increasing the photosynthetic efficiency and production of biomass. It becomes mandatory to invent techniques for optimizing biofuel production and also to provide environmentally friendly fuels which are now possible with the utilization of biotechnology techniques. These methods also counteract with the global warming and efficiently fight with the problem of environmental pollution implicated by the greenhouse gases (Delangiz et al. 2019). Thus, biofuel industries are a very important model of sustainability development in biotechnology. As they have conceptualized to be the alternative for the fossil fuels, scientists and researchers have to overcome the barriers in setting up one such industry or be a part of it.

Starting from the process optimization to the fine-tuning of the various processes available to get the product out of the industry is really cumbersome. To be successful, the organizations must continuously be responsive to the changing forces of the business environment (Downes and Mui 2000) and strive for distinctively excellent quality through their effective and efficient business processes. To strike a balance between the realistic and the imaginative ideas put forth in any research-based idea is also challenging while setting up a biotech industry, especially those which work on the sustainability sector. Business creation from biotechnology could be done through curriculum setting, financial and other policy incentives, and using their convening power to start-ups and shorten the learning curve (Hohnen 2017).

11 Summary

The process of training young researchers is a key to ensure the feasibility of these start-ups in the long term; however, many universities have curricula that are far from generating conditions that favour the development of new businesses. In one of the recent articles in the European journal of innovation management, there was a statement which indicates that the driven passion within oneself and the ability to collaborate with peers and participate in the creation of new ventures in life science start-ups have always overshadowed the downside risks and the lower level of economic compensation. Passion-driven and motivation-filled work atmosphere would also be a fun-filled one. The overwhelming incentives and lucrative earnings of the other industries would pose a threat to biotechnology entrepreneurs at the first, but the passion within themselves would definitely allow them to choose a career in a start-up and to make a mark rather than earning high-rise figures from a venture capital industry.

References

Assa F, Jafarizadeh-Malmiri H, Ajamein H, Anarjan N, Vaghari H, Sayyar Z, Berenjian A (2016) A biotechnological perspective on the application of iron oxide nanoparticles. Nano Res 9(8):2203–2225

Cardon MS, Kirk CP (2015) Entrepreneurial passion as mediator of the self-efficacy to persistence relationship. Entrepreneurship Theory Pract 39(5):1027–1050

Chaudhary G, Singh SK (2019) Global status of genetically modified crops and its commercialization. In: Khoobchandani M, Saxena A (eds) Biotechnology products in everyday life. EcoProduction (Environmental Issues in Logistics and Manufacturing). Springer, Cham

Chok J, Sun Q (2007) Determinants of idiosyncratic volatility for biotech IPO firms. Financ Manag 36(4):107–122

Delangiz N, Varjovi MB, Lajayer BA et al (2019) Arab J Geosci 12:174

Dettenhofer M, Ondrejovič M, Slavica A, Kurtanjek Ž, Tapaloaga D, Rodian Tapaloaga P, Kapur Pojskić L, Durmić-Pašić A, Begović J, Nedović V, Dundar M, Gartland KMA, Miertuš S (2018) Current state and prospects of biotechnology in Central and Eastern European countries. Part II: new and pre-accession EU countries (CRO, RO, B&H, SRB). Crit Rev Biotechnol 29:1–19

Dilli S, Elert N, Herrmann AM (2018) Small Bus Econ 51:293

Downes L, Mui C (2000) Unleashing the killer application: digital strategies for market dominance. Harvard Business School, Boston

Evens RP, Kaitin KI (2018) The biotechnology innovation machine: a source of intelligent biopharmaceuticals for the pharma industry—mapping biotechnology's success. Clin Pharmacol Ther 95(5):528–32; Exp Rev Pharmacoecon Outcomes Res

Giones F, Brem A (2017) From toys to tools: the co-evolution of technological and entrepreneurial developments in the drone industry. Bus Horizons 60(6):875–884

Hager Y (2019) Medical communications: the "write" career path for you? Cold Spring Harb Perspect Biol 11(1)

Hohnen P (2017) A winning formula for sustainable chemical startups. Innov Forum 1

Horvath B, Khazami N, Ymeri P, Fogarassy C (2019) Investigating the current business model innovation trends in the biotechnology industry. J Bus Econ Manage 20(1):63–85

Joshi T (2018) The dynamics of knowledge sharing in the biotechnology industry: an Indian perspective technology innovation management review 8(1)

Kafarski P (2012) Rainbow code of biotechnology. Chemik 66(8):811–816

Kaitin KI (2010) Deconstructing the drug development process: the new face of innovation. Clin Pharmacol Ther 87(3):356–361

Katz F, Glass RI (2019) Mentorship training is essential to advancing global health research. Am J Trop Med Hyg 100(1_Suppl):1–2. https://doi.org/10.4269/ajtmh.18-0694. Accessed 14 Nov 2018

Khanna I (2012) Drug discovery in pharmaceutical industry: productivity challenges and trends. Drug Discov Today 17

Khilji S, Mroczkowski T, Bernstein B (2006) From invention to innovation: toward developing an integrated innovation model for biotech firms. J Prod Innov Manag

Kleter GA, Kuiper HA, Kok EJ (2019) Gene-edited crops: towards a harmonized safety assessment. Trends Biotechnol

Lamb JA, Curtin JA (2019) Translational medicine: insights from interdisciplinary graduate research training. Trends Biotechnol 37(3):227–230

Lastres HMM, Sarita A (1999) Informacao e Globalizacao na Era do Conhecimento. Rio de Janeiro, Editora Campus Ltda

Maak TG, Wylie JD (2016) Medical device regulation: a comparison of the United States and the European Union. J Am Acad Orthop Surg 24(8):537–543

Meshram A, Singhal G, Bhagyawant SS, Srivastava N (2019) Chapter 28—plant-derived enzymes: a treasure for food biotechnology. In: Enzymes in food biotechnology production, applications, and future prospects, pp 483–502

Moustafa K, Makhzoum A, Trémouillaux-Guiller J (2016) Molecular farming on rescue of pharma industry for next generations. Crit Rev Biotechnol 36(5):840–850

Munos B (2009) Lessons from 60 years of pharmaceutical innovation. Nat Rev Drug Discov 8:959–968

Munshi A (2019) Assessment of competitiveness of food-tech start-ups in India. Indian J Public Admin 65(1):201–224

Nagendran M, Riordan DP, Harbury PB, Desai TJ (2018) Automated cell-type classification in intact tissues by single-cell molecular profiling. eLife 7

Ninawe (2019) Accelerating research and education for the growth of life sciences and biotechnology. Int J Biol 4(1):12–22

Octave S, Thomas D (2009) Biorefinery: toward an industrial metabolism. Biochimie 91(6):659–664

Pessoa MG, Vespermann KAC, Paulino BN, Barcelos MCS, Pastore GM, Molina G (2019) Newly isolated microorganisms with potential application in biotechnology. Biotechnol Adv 37(2):319–339

Rohela GK, Srinivasulu Y, Rathore MS (2019) A review paper on recent trends in bio-nanotechnology: implications and potentials. Nanosci Nanotechnol Asia 9(1):12–20(9)

Segers JP (2015) J Glob Entrepr Res 5:16

Stuart TE, Ding WW (2006) When do scientists become entrepreneurs? the social structural antecedents of commercial activity in the academic life sciences. Am J Sociol 112(1):97–144

Stuart TE, Ozdemir SZ, Ding WW (2007) Vertical alliance networks: the case of university—biotechnology—pharmaceutical alliance chains. Res Policy 36(4):477–498

Tohidyan Far S, Rezaei-Moghaddam K (2019) J Glob Entrepr Res 9:23

VijayPaul PE, Sangeetha V, Deepika RG (2019) Chapter 9—emerging trends in the industrial production of chemical products by microorganisms. Recent Develop Appl Microbiol Biochem 107–125

Xu N, Wei L, Liu J (2019) Recent advances in the applications of promoter engineering for the optimization of metabolite biosynthesis. World J Microbiol Biotechnol 35:33

Zahra SA (2007) Contextualizing theory building in entrepreneurship research. J Bus Ventur 22(3):443–452

Zeng AP (2019) New bioproduction systems for chemicals and fuels: needs and new development. Biotechnol Adv S0734-9750(19)30003-5

Gaurav Verma works at Clinical Research Center, Lunds University, Malmo, Sweden, to advance the knowledge of diabetes and to explore the mechanisms to treat the diabetic individuals. He is involved to identify a novel or unknown process through which T2D progresses and emphasizes on the development of novel therapies to reverse diabetes.

Srividhya Ravichandran did her research work on the effect of polyphenols on the oxidative stress mediated neuronal alterations in ageing rat brain. Currently, she is working as a technical staff at IIT Madras. Her passion towards science education drives her to be a scientific writer as well.

Some Success Stories and Products

Products of Biotechnology: The Out-Turn of Research and Production

Arpita Saxena

Abstract This chapter covers information about various products of biotechnology in the market already, or about to hit the market. The prehistoric, historic and modern findings in biological and biotechnological research have led to harnessing the capabilities of basic unit of life, i.e., the cell and its internal machinery. This is closely responsible for the advancements in the fields of microbiology, physiology, medicine, biochemistry and environmental studies; thereby, leading to a rapid amplification of tools and techniques of recombinant DNA technology, cell fusion techniques, tissue culture methods, genetic probes and bio-markers. All this has emerged as modern biotechnology, which has closely bound research with industry and business. These fundamental discoveries need to be translated into applications within a fixed or predictable time frame. Although conversion of scientific concepts into ideas, ideas into proof of concept (POC) and POC into finished products is not an easy task and it takes a long time, immense efforts and lot of financial investments to complete this journey, still a lot of entrepreneurs have been able to accomplish this and have made a big difference in the socio-economic lives of commoners.

Keywords Products · POC · Business · Bioprocessing · Enzymes

1 Introduction

All research should ideally lead to business and all business leads to products. These products can be in the form of goods or services. Some products are designed for consumption by common people directly, known as consumer products and some for serving various business levels like producers, who further produce other products out of these primary products or retailers, etc. Biotechnology industry is generally characterized by frequently changing business models pertaining to the innovations that keep the sector enormously diverse and flexible. Although this flexibility also brings along uncertainty which is an indispensable part of biotechnology business

A. Saxena (✉)
AAPT Foundation, SPAK megAcorp, S-9, Opp. DKMM Homeopathic Medical College, Aurangabad 431004, India
e-mail: arpitaspak@gmail.com

A. Saxena (ed.), *Biotechnology Business - Concept to Delivery*, EcoProduction,
https://doi.org/10.1007/978-3-030-36130-3_10

(Konde 2009). Market economy is one of the fundamental driving forces that drive this industry; other forces being social, political, environmental demands of the world population, to continue driving this industry further (Soetaert and Vandamme 2006). This industrial revolution in the field of biotechnology has led to product-oriented research. This is a modern wave which has surpassed the earlier two waves, namely medical or red biotechnology and agricultural or green biotechnology (Singh 2014). In the Indian scenario, the government has decided to invest US$5 billion to develop infrastructure and research laboratories so as to achieve a growth up to US$100 billion by 2025. Presently, the global share of Indian biotech industry is about 2% growing at a compound annual growth rate (CAGR) of 20% (Next what business, Accessed 2019).

2 General Products of Biotechnology

The traditional methods of food preservation have a long history dating back to the discovery of fermentation, brewing, pickling and so on. These included microbes and their use which was well explored since a long time. These methods led to processing of certain range and variety of products. Followed by this, events like bioprocessing and enzymatic process yielding, led to production of many useful products with large market such as amino acids, organic solvents, acids, vaccines, etc. (Demain 2007). Today these methods have been industrially exploited and have been developed rapidly from the business point of view. There are specific broad sectors of biotechnology industry like **food, pharmaceuticals, medical, agriculture and environment**. I shall try to address some of them separately in this chapter. There are also some general consumer products besides the above-mentioned categories, which are manufactured using biotechnological processes such as the following (Biotechnology Industry organization 2019).

- Cosmetics and personal care items—biotechnologically textured products are highly pure, non-irritating, smoother, less greasy, enhanced hair-care properties and environmentally sustainable.
- Bread—genetically modified microorganisms with enzymes to ensure better quality bread having longer shelf life and eliminates the carcinogen potassium bromate.
- Vitamin B2—genetically enhanced microbes produce this in one step fermentation reducing 33% CO_2 emission and use of energy.
- Diapers—using bacillus a biodegradable polymer is produced which is ecofriendly and skin friendly as well.
- Detergent—biotech enzymes like proteases, amylases and lipases help cleaning clothes brighter and save energy.
- Tissue paper—using wood bleaching enzymes the process of pulping is faster, more environmentally friendly and cost effective.

- Textiles and stonewashed jeans—use of biotech cellulose creates fabrics that dye better and are soft.
- Foam and nylon—The biotech processed foam is used to make furniture and nylon to make carpets of softer better quality fiber. This is more durable, stain and UV resistant than traditionally manufactured nylon.
- Plastics—biodegradable polymer, manufactured using bacillus microbe are very useful in packaging of food and beverages, in making food service wares and containers and are ecofriendly.
- Synthetic rubber—natural substances are polymerized to synthetic rubber and other elastomers. These are high in purity and low in cost.

Apart from these, there are some more products (Table 1) that are either about to hit the market floor or are in the market already. These products have started making their customer base and their specific niche in the market is rapidly expanding.

3 Pharmaceutical Products

A usual product of 'drug like' capabilities on an average, takes ten years to find its place in the market. It starts from virtual screening of millions of molecules through tools of bio-informatics. This in silico study not only facilitates screening, refinement, characterization of side effects and resistance of drug candidate, but also profiles a high throughput data related to its genome, epigenetics, proteomics and ribosome-related information (Xia 2017). The in vitro study follows this virtual screening process and on a daily basis, and it is capable of generating data on hundreds and thousands of compounds (Blass 2015). This is followed by mechanistic study along with the clinical studies to examine the dosage regimen, efficacy, effects and side effects, pharmacokinetics and quality control for the drug entity (Aronson et al. 2018). The road to make a biotech drug is filled with a lot of hurdles and U-turns and the probability of failure haunts the scientists on every step. As per Professor Gary P. Pisano, a renowned business analyst, it is critically important to get the right people to work as he says, "Once you attract bright, young scientists, you need to get them to realize this is not a postdoc lab, this is a business. It's not enough just to do the experiment; somebody's got to take the ball and run with it" (Hanna 2000). Most biotech drugs can be tailored and the patients can be treated more effectively and even the genetic makeup of the patient can be changed depending upon the case to case basis. Pisano further adds that "The $1 and $2 billion drugs will give way to $200–$300 million drugs. That will be a very different world for big drug companies, with different cost structures and resource-allocation processes" (Hanna 2000). Experts say that the process of drug discovery is a complex process and the genome revelation will facilitate the identification of molecular interactions for a drug. Listed in Table 2 are some biopharmaceuticals in global market and their manufacturers from around the world along with their uses and applications.

Table 1 Some general biotech products (Sotiriadis 2017)

S. no.	Product name	Manufacturing company/laboratory	Advantages
1.	Drinks made with algae (Grebow 2017)	LifeDNA TerraVia	Vegetarian DHA and vegetarian Omega-3 Algal protein and lipid powder
2.	Aquamin	Stauber	Aquamin derived from *Lithothamnion* red algae and is suited for everything from bakery products to confectionery, snacks, smoothies, fruit juices, and vegetable drinks and is rich in calcium and trace minerals
3.	Synthetic spider silk (Matchar 2017)	University of Cambridge	It is found to be stronger than steel and tougher than Kevlar making it one of the sturdiest materials found in nature. It can be stretched several times its length before it breaks
4.	Mushtari (Oh 2015)	Mediated matter group and stratasys	It is a 3D-printed wearable that can convert sunlight into usable products like sucrose, desired pigments, drugs, food, fuel or scents
5.	Cow-free cow milk	Perfect day	Contains same proteins like an actual cow minus lactose Can be produced more locally and sustainably Shelf life of six months so storage and logistics becomes easy
6.	Lab-grown meat	Memphis Meats	Cheap, scalable, no animal suffering, low pressure on environment, programmable food and meat produced molecule by molecule

(continued)

Table 1 (continued)

S. no.	Product name	Manufacturing company/laboratory	Advantages
7.	Cell free	Cell-free Tech	User-friendly easy to use kits containing plasmids carrying the genetic software instructions like an app. This eliminates the need for microbial culturing, training and equipment and customized products can be obtained
8.	MicrobeMiner	Prospective research	It mimics the stimuli present in soil to activate the silent operons of *Streptomyces* bacteria thereby producing variety of antibiotics
9.	Mushroom lamp	Ecovative	Eco-lamps, biodegradable, sustainable, decorative and turns down plastic pollution

4 Food Products from Industrial Biotechnology

Food is the basis of sustenance and food processing has made food a way of recreational activity as well. Socialization over food is a common thing since oldest times and these days, it has become just more significant. Keeping this in mind, the food sector has rapidly evolved and significant business associations have formed all over the world liaisoning modern biotechnology research skills with large-scale manufacturers and marketers of food products (Lawrence 1988). Biotechnology, which was earlier only being looked upon as medical and pharmaceutical revolution, now is making a billion-dollar market in food as well. Some of the food and beverages products are discussed (Maud Kordylas 1992) in Table 3.

5 Products of Agricultural Biotechnology

For a very long time research in agricultural biotechnology has been focussed on improving crops yield, crop nutritional value and protection of crops against diseases, pests and natural calamities. This has led to a variety of products with higher yield, insect resistance, disease resistance, altered nutritional profile, enhanced shelf life, draught tolerance, herbicide tolerance and many more specific traits desired by agriculturists. There have been debates around the world about safety of biotech crops, still biotech crops have been cultivated for nearly two decades and consumed

Table 2 Biopharmaceutical products (Konde 2009; Demain 2007; Singh 2014; Biotechnology Industry Organization (BIO), 2003)

S. no.	Biopharmaceuticals	Applications	Manufacturing company/companies
1.	Plasma derivatives, monoclonals, hormones, antifungals, anesthetics, diagnostic products	Medical and clinical applications	Bharat Serums and Vaccines, Mumbai
2.	Aranesp™ (darbepoietin alfa)	Anemia induced due to Chemotherapy in patients with non-myeloid malignancies	Amgen
3.	Human growth hormone	Growth deficiency	Genetech, Pharmacia
4.	Interferon-α	cancer, hepatitis	Schering-Plough, Roche, Serono, Schering AG
5.	Interferon-β	Multiple sclerosis, hepatitis	Biogen, Serono, Schering AG, Chiron
6.	Recombinant drugs, cardiovascular diseases, vaccines	Heart diseases and vaccines for various diseases	Bharat Biotech International, Hyderabad
7.	Human insulin	Diabetes	Novo Nordisk, Eli Lilly
8.	Argatroban	Anticoagulant for use in patients with or at risk of heparin-induced thrombocytopenia undergoing percutaneous coronary interventions	Texas Biotechnology Corp. and GlaxoSmithKline
9.	Tissue plasminogen activator	Blood clot	Genentech
10.	BOTOX ® COSMETIC (botulinum toxin type A)	Temporary improvement in appearance of moderate to severe frown lines in adults 65 or younger	Allergan Inc.
11.	Pediatric and childhood vaccines, DNA-based vaccines	Animal- and human-health products	Indian Immunologicals, Hyderabad
12.	FOLLISTIM™ (follitropin-beta)	Treatment of primary and secondary hypo-gonadotropic hypogonadism in men	Organon (unit of Akzo Nobel)
13.	Polyvalent vaccines, drugs/formulations	Human indications including pain management, diabetes and renal disease	Panacea Biotec, New Delhi
14.	Granulocyte-colony stimulating factor (G-CSF)	Neutropenia/White blood cell	Amgen, Roche, Schering

(continued)

Table 2 (continued)

S. no.	Biopharmaceuticals	Applications	Manufacturing company/companies
15.	Factor VIII	Hemophilia	Bayer
16.	Rituximab	Non-Hodgkin's lymphoma	Genentech/Idec
17.	INTEGRA ® dermal regeneration template	Repair of scar contractures	Integra Life Sciences Holding Corp and Ethicon Inc.
18.	Glucocerebroidase	Gaucher's disease	Genzyme
19.	OLUX ® Foam (clobetasol proprionate)	Short-term topical treatment of mild to moderate plaque-type psoriasis of non-scalp regions excluding the face and intertriginous areas	Connetics Corp.
20.	Blood plasma proteins, recombinant proteins		Reliance Life Sciences, Mumbai
21.	Erythropoietin	Anemia	Amgen, Johnson & Johnson, Roche, Kirin, Sankyo
22.	Therapeutic antibodies	Cancer	Glaxo, Amgen, Genentech
23.	SecreFlo™ (synthetic porcine secretin)	Aid in location and cannulation of pancreatic ducts in patients undergoing endoscopic retrograde cholangiopancreatography	Repligen Corp.
24.	Etanercept	Rheumatoid arthritis	Amgen, Wyeth
25.	Xyrem ® (sodium oxybate)	Cataplexy associated with narcolepsy	Orphan Medical Inc.
26	Infliximab	Crohn's disease	Johnson & Johnson
27.	Orfadin ® (nitisinone)	Hereditary tyrosinemia type 1	Swedish Orphan International AB and Rare Disease Therapeutics Inc.
28.	Trastuzumab	Breast cancer	Roche
29.	Gleevec™ (imatinib mesylate)	Gastrointestinal tumors and Philadelphia chromosome-positive chronic myeloid leukemia	Novartis AG
30	Palivizumab	Prevention against respiratory syncytial virus	Medimmune

Table 3 Food products obtained by using biotechnological processing (FAO International Technical Conference 2010)

S. no.	Products	Country/Company	Benefits	Process of production
1.	Rob	Sudan	Butter extraction	Fermentation (Dirar 1992)
2.	Merissa, Opaque beer	Sudan	Local beer	5% sorghum malt is used along with a caramelized sorghum product called surji (Dirar 1976, 1978)
3.	Fermented soy sauce	Thailand, Japan, China	Economic gain for the soy sauce industry and greater value added to the product in terms of quality and safety	Using starter culture technology and the use of improved bioreactor technology. *Aspergillus oryzae* and *Saccharomyces rouxii* are used as microbes for culturing and fermentation (Valyasevi and Rolle 2002)
4.	Nham (fermented pork sausage)	Thailand, Viet Nam, Lao and Combodia	Traditionally produced nham is considered high risk by Thai health authorities, while biotechnologically processed Nham was endorsed for its safety and quality	Acidity of the culture was rapidly increased that prevented the growth of bacterial pathogens. Nitrite was checked and monitored. An innovative pH indicator was used to mark end point of fermentation. RAPD markers were used for the molecular typing of approximately 100 bacterial strains (Paukatong and Kunawasen 2001)

(continued)

Table 3 (continued)

S. no.	Products	Country/Company	Benefits	Process of production
5.	Som Fug (Fermented fish paste)	South east Asian countries	High discriminatory power of biotechnology in differentiating lactic acid bacteria at the strain level	*Lactobacillus pentosus* and *Lact. Plantarum* are the garlic fermenting lactic acid bacteria associated with Som Fug fermentations (Paludan-Muller et al. 2002)
6.	Flavor production from alkaline-fermented beans	African Savannah region	High-value compounds such as flavor compounds are derived from low investment fermentation	Alkaline fermentation of the African locust bean, dawadawa (Steinkraus 1995)
7.	Vanillin, Monascin		Flavoring and coloring pigments	

by billions of people around the world. Scientists believe that to improvise the crops, traditional breeding methods have the same consequences as using biotechnological techniques. An interesting application of selective alteration of specific genes is that specific allergens in particular crops can be checked and a non-allergenic variety of the same crop can be developed targeting the consumer group those are allergic to specific type of foods. Generally speaking, the scientists aim to check hunger and malnutrition through gene alterations and specifically bringing up varieties of foods with particular desired traits. Some biotechnologically grown crops have been enlisted in Table 4 along with their advantages over the previous traditional crops.

6 Summary

Apart from the crops that are genetically modified for achieving specific traits, there are other applications of agricultural biotechnology research that has led to products of high commercial value like vaccines, antibiotics and nutritional supplements. Genetically modified foods may be prepared with specific antigenic proteins that trigger immune response in consumers and/or boost their resistance to pathogens. The antibiotics obtained through biotech applications in plants reduce the cost of production, risk of contamination and production time. This book includes a chapter on nutraceuticals as they are a major player in economic growth and commercial value addition of food and pharma products. The beta carotene-rich golden rice is a transgenic grain rich in Vitamin A and certain other nutrients, which is valuable

Table 4 GMO crops and their advantages (Young 2017; ISAAA 2019)

S. no.	Crops/GMOs	Plantation areas	Advantages
1.	Maize	USA, South Africa, Argentina, Brazil, Canada, Paraguay, Colombia, EU, Philippines, Vietnam, Uruguay, Cuba, Honduras	85% maize grown in the USA is genetically modified. Herbicide-tolerant and insect-resistant varieties available
2.	Soy	USA, South Africa, Argentina, Bolivia, Brazil, Canada, Paraguay, Mexico, Chile, Costa Rica	Most economically relevant crop. Herbicide-tolerant and insect-resistant varieties available, another variety available is with high oleic acid content
3.	Canola	Canada, USA	Herbicide-tolerant gene containing variety is available along with a high Laureate variety, which is used for chocolate candy coatings, icings, toppings, coffee whiteners and few cosmetics
4.	Tomato	China and USA	These tomatoes contain genes for delayed ripening that provide longer shelf life, higher commercial value and better taste
5.	Alfalfa	Canada, Mexico, USA	Herbicide-tolerant variety contain genes that provide resistance to broad-spectrum herbicides thus reducing crop injury
6.	Potato	USA and Canada	Insect-resistant variety is resistant to pests and insects. The virus-resistant and low acrylamide varieties are also available
7.	Rice	USA	Herbicide-tolerant variety provides protection against broad spectrum and environmentally benign herbicides
8.	Cotton	USA, South Africa, Australia, Argentina, Brazil, Costa Rica, Colombia, Mexico, Paraguay	Herbicide-tolerant and insect-resistant varieties available
9.	Papaya	China and USA	There is a virus-resistant variety that provides in-built protection against Papaya Ringspot virus (PRSV)
10.	Sugar beet	USA and Canada	A herbicide-tolerant gene facilitates the growers with reduced number of cultivations

to check dietary deficiencies and resist diseases. Applications of agro-biotech in producing specific nutrient-rich crops are classic example of success of scientific intervention over traditional methods.

There are success stories of start-ups that transformed into Pvt. Ltd companies like Advanced Biotech Products (http://www.advanced-biotech.com/), Chennai. They are affiliated to Encoll Corp., a USA-based company, and have patented a technology to manufacture highly purified collagen. This collagen is surface modified using collagen-phosphorylation technology to make tissue regenerative and remodeling products for clinical and surgical purposes. They provide products for dermal surgical, general surgical, dental surgical and skincare cosmetics. They are working to achieve gene therapy, stem-cell delivery and target-based cure for diseases, using their technology base.

There are many companies that manufacture self-test kits to monitor health and diagnose diseases/pregnancy cases. Just like the popular pregnancy testing kit, there are ovulation detector kits, HIV and STD test kits and many more options for consumers to personally diagnose these conditions. Forensic DNA analysis tools, lie-detecting equipment and biometric capturing centers are another category of products/services those have a demand in the market. There is a fairly big market for such products.

Energy sector remains a major investment platform as society is transforming from oil economy to bioeconomy. Bio-diesel, bio-ethanol and bio-oils are being produced from feedstocks of good quality. Microbial strains are being engineered for large spectrum applications. There are many companies that are into culture, genetic modification and production of such microbes like Aequor Inc. that engineers marine microbe that kills foul smell and biofilm over the water body; or Chain Biotechnology Ltd. that makes microbial hosts for engineering anaerobic bacteria. Similarly, Greenlight Biosciences facilitates cell-free bioprocessing technology, while Synthetic Genomics with its advanced genetic engineering techniques can produce microbial cell lines, DHA Omega-3 and Astaxanthin.

Indian biotechnology industry was primarily into manufacturing and production of biopharmaceuticals with a USA-based company as a technology partner. Gradually the scene has changed and there are many biotech start-ups mushrooming up with proprietary technology and innovation. The government support is also very favorable for start-up culture with many policies and funding support schemes for new entrepreneurial ideas. Education system itself has started to accommodate entrepreneurial skills in the curriculum which represents a positive change.

7 Conclusion

Research has moved from microbial fermentation related enzymes to DNA and RNA modifications and customized production of desired entities. With this transformation, a horizon has opened up for stakeholders of new biotech products and services. There are scientific and academic communities collaborating with each other to match up with the market demand and speed. Specific categories of products from pharmaceutical, food and agricultural sectors are always flowing into market dynamically while offbeat general products are matching up with these categories building up a niche. As the education and awareness in the society increases, acceptance of these products will increase resulting in the vertical mobility of biotech products on the ladder of new innovations and inventions.

Acknowledgements I thank my mentor Dr. A. K. Saxena.

References

Aronson JK, La Caze A, Kelly MP, Parkkinen VP, Williamson J (2018) The use of mechanistic evidence in drug approval. J Eval Clin Pract 24(5):1166–1176. https://doi.org/10.1111/jep.12960. Accessed 11 Jun 2018

Bio Biotechnology Industry organisation 'Consumer Products Made with Industrial Biotechnology'. https://www.bio.org/articles/consumer-products-made-industrial-biotechnology. Accessed 21 July 2019

Biotechnology Industry Organization (BIO), 2003 FDA Approves 20 New Biotech Products, 15 New Indications. Accessed 17 July 2019

Blass BE (2015) Chapter 4—In vitro screening. In: Basic principles of drug discovery and development, pp 143–202. https://doi.org/10.1016/B978-0-12-411508-8.00004-9

Demain AL (2007) The business of biotechnology. Ind Biotechnol, Gen publishing inc., a mary ann liebert inc. company 3(3)

Dirar HA (1976) The art and science of *merissa* fermentation. Sudan Notes Rec 57:115–129

Dirar HA (1978) A microbiological study of Sudanese *merissa* brewing. J Food Sci 43:1683–1686

Dirar HA (1992) Sudan's fermented food heritage applications of biotechnology to traditional fermented foods, National Academies Press (US). https://www.ncbi.nlm.nih.gov/books/NBK234677/

FAO International Technical Conference (2010) Agricultural biotechnologies in developing countries: options and opportunities in crops, forestry, livestock, fisheries and agro-industry to face the challenges of food insecurity and climate change (ABDC-10) Guadalajara, Mexico, 1–4 Mar 2010 Current status and options for biotechnologies in food processing and in food safety in developing countries

Grebow J (2017) Algae ingredients turn heads in food and drinks. In: Nutritional outlook. http://www.nutritionaloutlook.com/food-beverage/algae-ingredients-turn-heads-food-and-drinks. Accessed 16 July 2019

Hanna J (2000) The business of biotech. https://hbswk.hbs.edu/item/the-business-of-biotech. Accessed 20 July 2019

ISAAA Pocket K No. 2: Plant products of biotechnology. https://www.isaaa.org/resources/publications/pocketk/2/default.asp. Accessed 8 July 2019

Konde V (2009) Biotechnology business models: an Indian perspective. J Commer Biotechnol 15:215–226. https://doi.org/10.1057/jcb.2008.52

Lawrence RH (1988) New applications of biotechnology in the food industry. In: Biotechnology and the food supply: proceedings of a symposium national research council (US) commission on life sciences. National Academies Press (US), Washington (DC)

Matchar E (2017) New artificial spider silk: stronger than steel and 98 percent water. https://www.smithsonianmag.com/innovation/new-artificial-spider-silk-stronger-steel-and-98-percent-water-180964176/. Accessed 16 July 2019

Maud Kordylas J (1992) Biotechnology for production of fruits, wines, and alcohol applications of biotechnology to traditional fermented foods, National Academies Press (US). https://www.ncbi.nlm.nih.gov/books/NBK234683/
Next what business; '10 Profitable Biotechnology Business Ideas' https://nextwhatbusiness.com/biotechnology-business-ideas/. Accessed 20 July 2019
Oh E (2015) 'Neri Oxman's "Mushtari" is a 3D printed wearable that makes products from sunlight'. https://www.archdaily.com/769818/neri-oxmans-mushtari-is-a-3d-printed-wearable-that-makes-products-from-sunlight. Accessed 16 JUly 2019
Paludan-Muller C, Valyasevi R, Huss HH, Gram L (2002) Genotypic and phenotypic characterization of garlic-fermenting lactic acid bacteria isolated from Som–Fak, a Thai low–salt fermented fish product. J Appl Microbiol 92:307–314
Paukatong KV, Kunawasen S (2001) The hazard analysis and critical control point (HACCP) generic model for the production of Thai fermented pork sausage (Nham). Berl Münch Tierärztl Wschr 114:327–330
Singh RS (2014) Industrial biotechnology: an overview. In: Singh RS, Pandey A, Larroche C (eds) Advances in industrial biotechnology. IK International Publishing House Pvt. Ltd., India, pp 1–35
Soetaert W, Vandamme E (2006) The impact of industrial biotechnology. Biotechnol J 1:756–769
Sotiriadis E (2017) 5 wild biotech products that will touch our lives in the near future. https://singularityhub.com/2017/11/24/5-wild-biotech-products-of-the-future/. Accessed 6 July 2019
Steinkraus KH (1995) In: Steinkraus KH (ed) Indigenous fermented foods involving an alkaline fermentation, 2nd edn. NY, Marcel Dekker, pp 349–362
Valyasevi R, Rolle RS (2002) An overview of small-scale food fermentation technologies in developing countries with special reference to Thailand: scope of their improvement. Int J Food Microbiol 75:231–239
Xia X (2017) Bioinformatics and drug discovery. Curr Topics Med Chem 17(15):1709–1726. https://doi.org/10.2174/1568026617666161116143440
Young C (2017) 7 most common genetically modified foods. https://www.huffpost.com/entry/diet-and-nutrition_b_4323937. Accessed 27 Aug 2019

Arpita Saxena is presently self employed at SPAK megAcorp, her proprietary venture. Arpita was awarded her PhD in Biotechnology in 2012 off campus from Guru Nanak Dev University, Amritsar, India while working at the department of Cancer Pharmacology, Indian Institute of Integrative Medicine (CSIR), Jammu, India. During PhD her work was focused on understanding the methods to evaluate cell death, apoptosis and the mechanisms underlying apoptosis.

The manufacturer's authorised representative in the EU is Springer Nature Customer Service Centre GmbH, Europaplatz 3, 69115 Heidelberg, Germany. If you have any concerns regarding our products, please contact ProductSafety@springernature.com

Printed and bound by CPI Group (UK) Ltd, Croydon, CR0 4YY
22/09/2026
02220948-0013